SOLUBILITY OF POLYSACCHARIDES

Edited by **Zhenbo Xu**

Solubility of Polysaccharides
http://dx.doi.org/10.5772/66033
Edited by Zhenbo Xu

Contributors

Anna Ptaszek, Joanna Kruk, Michał Pancerz, W. M. C. Sameera, Bimali Jayawardena, Dinesh Raveendra Pandithavidana, Erich Von Borries-Medrano, Mónica Jaime Fonseca, Miguel Ángel Aguilar-Méndez, Qingbin Guo, Changlu Wang, Xinzhong Hu, Shao-Chi Wu, Zhenbo Xu, Jagadish Chandra Roy, Fabien Salaün, Stéphane Giraud, Ada Ferri, Guan Jinping, Yan Chen

Published by InTech
Janeza Trdine 9, 51000 Rijeka, Croatia

Publishing Process Manager Dajana Pemac
Technical Editor SPi Global
Cover InTech Design team

First published November, 2017
Printed in Croatia
Legal deposit, Croatia: National and University Library in Zagreb

Additional hard copies can be obtained from orders@intechopen.com

Solubility of Polysaccharides, Edited by Zhenbo Xu
p. cm.
Print ISBN 978-953-51-3649-1
Online ISBN 978-953-51-3650-7

For EU product safety concerns:
IN TECH d.o.o., Prolaz Marije Krucifikse Kozulić 3, 51000 Rijeka, Croatia,
info@intechopen.com or visit our website at intechopen.com.

Meet the editor

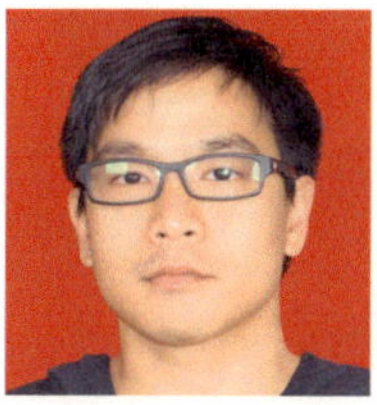

Zhenbo Xu, an associate professor, School of Food Science and Engineering, South China University of Technology (SCUT), grew up in China and received his bachelor's degree from SCUT in 2005. He joined Professor Mark E. Shirtliff's Laboratories in the University of Maryland as a joint PhD student during 2009 to 2011 and obtained his PhD degree majoring in saccharide engineering from SCUT in 2011. Dr. Xu was appointed as an assistant professor at SCUT after completing his doctorate and began his investigations on food-borne pathogens and bioinformatics. In 2016, Dr. Xu was promoted to associate professor in SCUT and became an adjunct professor in the University of Maryland in 2017. In recent years, he had more than 100 publications in academic journals, including 'Clinical Microbiology and Infection', 'Journal of Clinical Microbiology', 'International Journal of Food Microbiology' and 'Scientific Reports'. He got more than 400 citation times totally, and the H-index is 17. So far, Dr. Xu has received various awards, with the most representative being the Pearl River S&T Nova Program of Guangzhou (2017) and National Outstanding PhD Dissertation (2011).

Contents

Preface

Essential for living organisms, polysaccharides are polymeric carbohydrate molecules composed of long chains of monosaccharide units bound together by glycosidic linkages and on hydrolysis release the constituent monosaccharides or oligosaccharides, including storage polysaccharides and structural polysaccharides.The polysaccharides usually consist of monosaccharides, like glucose, fructose and glyceraldehyde. Based on this, polysaccharides could be classified into homopolysaccharide (distinctive types) and heteropolysaccharides (different types). The solubility of different polysaccharides is apparently different, resulting from the structure of the polysaccharides, and this limits the application of the polysaccharides.

In this book, the principles on polysaccharides' solubility and structure, methodologies and application of polysaccharides have been reviewed. The chapters in this book include the relationship between structure and solubility of polysaccharide, the experimental and computational researches on polysaccharide solubility and the common polysaccharide, in the presence of which we could learn more on the solubility of polysaccharides, the research methods andthemodification method.

Zhenbo Xu
Associate professor
School of Food Science and Engineering
South China University of Technology (SCUT)
Guangzhou, China

Section 1

Introduction

Chapter 1

Introductory Chapter: Polysaccharides and their Solubility

Zhenbo Xu

Additional information is available at the end of the chapter

http://dx.doi.org/10.5772/intechopen.70137

Essential for living organisms, polysaccharides are polymeric carbohydrate molecules composed of long chains of monosaccharide units bound together by glycosidic linkages and on hydrolysis releases the constituent monosaccharides or oligosaccharides [1, 2], including storage polysaccharides (such as starch and glycogen) and structural polysaccharides (cellulose and chitin). Starch is universally existed in plants, with cellulose and chitin found in the cell walls of plants, fungi, and other organisms [3]. Based on the component units, polysaccharides are classified into homopolysaccharide (distinctive types) and heteropolysaccharides (different types). Unbranched polysaccharides contain only α-1, 4 linkages, however, branched polysaccharides compose molecular linkage via α-1, 4 and α-1, 6 glycosidic bonds [4–6]. The rate of the bonds formation may vary. The plant-based amylopectin and the animal-based glycogen contain a branch every 30 and 10 units, respectively. During digestion, as a major catalyst on the branched polysaccharides, α-amylase only digests α-1, 4 glycosidic bonds, and the rest disaccharide/polysaccharide fragments contains α-1, 6 bonds [7]. Homoglycans with two types of saccharide linkages or heteroglycans composed of two types of saccharide are more soluble than purely homogeneous polymers [5]. However, poor solubility of polysaccharides significantly limits its development and application.

In this book, the principles on polysaccharides solubility and structure, methodologies and application of polysaccharides have been reviewed. The five chapters in this book include polysaccharides structure and solubility, osmotic properties of polysaccharides solutions, polysaccharides in solution experimental and computational studies, antioxidative activity of sulfated seaweed polysaccharides by variety-assisted extraction and starch galactomannans mixtures rheological and viscosity behavior in aqueous systems for food modeling.

Author details

Zhenbo Xu

Address all correspondence to: zhenbo.xu@hotmail.com

South China University of Technology, China

References

[1] Fukaya Y, Sugimoto A, Ohno H. Superior solubility of polysaccharides in low viscosity, polar and halogen-free 1, 3.- dialkylimidazolium formats. Biomacromolecules. 2006; **7**:3295-3297

[2] Whistler RL. Solubility of polysaccharides and their behavior in solution. Advances in Chemistry. 1973;**117**:242-255

[3] Kardosova A, Machova E. Antioxidant activity of medicinal plant polysaccharides. Fitoterapia. 2006;**77**:367-373

[4] Gooneratne J, Magsak-Newman G, Robertson JA, et al. Investigation of factors that affect the solubility of dietary fiber, as nonstarch polysaccharides, in seed tissues of mung bean (*Vigna radiata*) and Black Gram (*Vigna mungo*). Journal of Agricultural and Food Chemistry. 1994;**42**:606-611

[5] Zakrzewska ME, Bogel-Lukasik E, Bogel-Lukasik R. Solubility of carbohydrates in ionic liquids. Energy and Fuels. 2010;**24**:737-745

[6] Fu G, Liu C, Tu Z. Studies on the isolation, purification and composition of water-solubility polysaccharides from *Agrocybe Chaxingu*. Food Science. 2005;**9**:151-155

[7] Naskar B, Dan A, Ghosh S, et al. Viscosity and solubility behavior of the polysaccharide inulin in water, water + dimethyl sulfoxide, and water + isopropanol media. Journal of Chemical & Engineering Data. 2010;**55**:2424-2427

Section 2

Solubility and Structure of Polysaccharides

Chapter 2

Polysaccharides: Structure and Solubility

Mark Q. Guo, Xinzhong Hu, Changlu Wang and Lianzhong Ai

Additional information is available at the end of the chapter

http://dx.doi.org/10.5772/intechopen.71570

Abstract

Understanding the solubility of polysaccharides is extremely important for their food applications as most functions of polysaccharides including stability, emulsifying property, drug delivery, membrane forming properties, etc., are all achieved in aqueous solution. This chapter aims specifically at the mechanism of solubility of polysaccharides from the molecular level. General understandings of the solubility including definition, testing methods, and the solution behaviors were provided; the relationships between polysaccharide solubility and the structural features in terms of molecular weight, degree of branching, charging properties, chain flexibility, and the special groups were all discussed. With all the information provided, the molecular modification and further applications of polysaccharides in both food and nonfood areas could be promoted.

Keywords: polysaccharides, structure, solubility, conformation, solution behavior

1. Introduction

Polysaccharides are polymeric carbohydrate molecules composed of long chains of monosaccharide units bound together by glycosidic bonds. Natural occurring polysaccharides exhibit distinct structural features in terms of molecular weight, monosaccharide composition, glycosidic linkage patterns, configuration (α or β), charging properties, degree of branching, etc. These diversified structural properties determine the functional properties of polysaccharides, such as solubility and rheological properties, which in turn benefit their extensive applications in both food and nonfood areas. As most polysaccharides perform their function in aqueous solution, understanding the solubility of polysaccharides, therefore, becomes critically important [1]. This chapter specifically addressed the mechanisms of polysaccharide solubility from molecular level. The relationships between polysaccharide solubility and molecular structures were established. It also should be noted that the current chapter only focused on the polysaccharides in aqueous solution; the solubility of polysaccharides in other organic solvents was not covered.

2. Solubility of polysaccharides

Polysaccharides display a wide range of solubility; some are water insoluble, e.g., cellulose; some are only hot water soluble, e.g., starch; and some are readily dissolved in cold water, such as pullulan and gum arabic. The dissolution of polysaccharides is different from that of the small crystalline molecules. The dissolution of most crystalline small molecules involved the disintegration of the crystalline structure and release of the separate atoms, ions, polysaccharides dissolution is a continuous hydration process with the conversion of inter-polysaccharide binding to polysaccharide-water binding, and most of the non-starch polysaccharides are in amorphous state. The dissolution process is more or less assisted by entropy as the molecules assume lower-energy conformations.

Polysaccharides have strong affinity to water molecules due to the presence of multi-OH groups. However, this also leads to a strong interaction among polysaccharide molecules via hydrogen bonding. Therefore, the balance between molecule-molecule interaction and molecule-water interaction is the key to understand the polysaccharide solubility. For soluble polysaccharides, the interactions between polysaccharide molecules and water molecules are energetically favorable, and the solvent creates a solvating envelope around the polymer chain, which keeps the polysaccharide molecules away from each other. For polysaccharides with poor water solubility, the intramolecular interactions between polymer segments dominate, leading to aggregation and eventually precipitation or gelation when ordered molecular structure, e.g., junction zone, is formed. In between, under specific conditions, the polymer-polymer interaction can precisely compensate the polymer-water interaction, which referred as *theta* condition. Under *theta* condition, the chain conformation is defined solely by bond angles and short-range interactions given by the hindrances to rotation about bonds and polymer coil dimensions [2].

Second virial coefficient (A_2), which describes the contribution of the pair-wise potential to the pressure of the gas, could reflect the polymer-water interaction. Good solvent, poor solvent, and *theta* condition can be indicated when the second virial coefficient is above, below, or equal to zero, respectively. In a real experiment, the second virial coefficient of the polysaccharides in aqueous solution can be determined by static light scattering using Zimm plot (**Figure 1**) based on the Eqs. (1–2) [3]:

$$Kc/R_\theta = 1/M_w + 1/3\left(R_g^2/M_w\right)q^2 + 2A_2c \tag{1}$$

where K is an optical contrast factor, c is the polymer concentration, R_θ is the Rayleigh ratio (normalized scattering intensity), and the scattering vector q is defined as

$$q = 4\pi n_0 sin(\theta/2)/\lambda_0 \tag{2}$$

with n_0 as the refractive index and λ_0 as the wavelength in a vacuum.

For example, A_2 of xyloglucans from flaxseed kernel cell wall was reported as 3.9, 0.9, and $-0.54\ 10^{-4}\ cm^3\ mol/g^2$ in 0.5 M NaOH solution, H_2O, and 0.1 M $NaNO_3$ solution, respectively, indicating xyloglucan prone to dissolve in 0.5 M NaOH than 0.1 M $NaNO_3$ [4]. 0.5 M NaOH

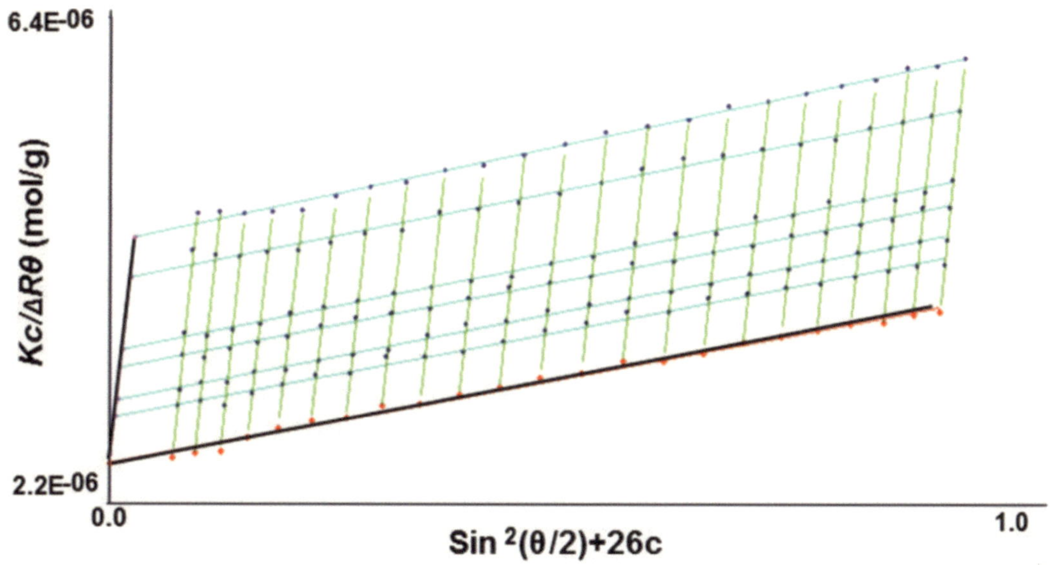

Figure 1. Zimm plot of polysaccharide from seed of Artemisia sphaerocephala Krasch determined by SLS (solvent, 0.5 M NaOH; temperature, 25°C). Adopted from Guo et al. [3].

has been reported as a good solvent for several different polysaccharides, such as β-glucan [5], arabino-glucuronoxylan isolated from seeds of *Artemisia sphaerocephala* Krasch [3], and gum ghatti [6]. This is because mild alkaline could break down the intermolecular hydrogen bonding, thus eliminating the aggregates in aqueous solution.

3. Structure and solubility relationships

The solubility of polysaccharides is determined by their molecular structures. Any structural feature that hinders the intermolecular association leads to higher solubility, such as branching structure, charged group (carboxylate group, sulfate, or phosphate groups); on the opposite, structural characters that promote the intermolecular association result in a poor solubility, such as linear chain, large molecular weight, and other regular structural characters.

3.1. Molecular weight and molecular weight distribution

Polysaccharides are polydisperse in molecular weight. Therefore the molecular weight of polysaccharides is mostly described in a statistic way, such as number average molecular weight (Mn), weight average molecular weight (Mw), and zeta average molecular weight (Mz), as shown in the below equations (Eqs. (3–5)). Here Ci refers to the concentration of molecules that having molecular weight of Mi. The molecular distribution of polysaccharides can be described by the polydispersity index (Eq. (6)). For monodispersed polymers, PDI = 1, while PDI less than 1.2 and above 2 are generally regarded as narrow disperse and wide disperse, respectively. Most natural occurring polysaccharides demonstrated high PDI value (above 2):

$$Mn = \frac{\sum C_i}{\sum C_i/M_i} \tag{3}$$

$$Mw = \frac{\sum M_i C_i}{\sum C_i} \tag{4}$$

$$Mz = \frac{\sum M_i^2 C_i}{\sum M_i C_i} \tag{5}$$

$$PDI = \frac{Mw}{Mn} \tag{6}$$

The molecular weight and molecular weight distribution play a critical role for the solubility of polysaccharides. High molecular weight molecules normally have a large excluded volume (Eq. (7)) that promotes intermolecular interaction of polysaccharide, and suppress its solubility. Almost all carbohydrate polymers with degrees of polymerization (DP) less than 20 are soluble in water [7]. Solubility decreases with the increase of molecular weight. For example, the amylose and amylopectin in starch are reluctant to dissolve in cold water due to high molecular weight, while maltodextrin (starch after chain cleavage by acid or enzyme) with the DP value less than 20 demonstrates very good solubility in cold water. The dissolution rate of polysaccharide samples is also highly affected by the molecular weight and molecular weight distribution. Higher molecular weight usually leads to lower dissolution rate, as disentanglement from the particle surface and subsequent diffusion to the bulk solution of large molecules take a longer time compared to that of small molecules. It has also been reported that samples with high polydispersity dissolved about twice as fast as monodisperse ones of the same Mn [1]:

$$V_h = \frac{[\eta]M}{\gamma N_A} \tag{7}$$

where V_h is the hydrodynamic volume, M is molar mass, N_A is Avogadro's number, $[\eta]$ is intrinsic viscosity, and γ is Simha's parameter related to the shape of a particle.

3.2. Charged polysaccharides vs. neutral polysaccharide

Charged polysaccharides are referred to polysaccharides that carry charged groups in the molecules, which include both negatively (acidic polysaccharides) and positively charged polysaccharides. The charged groups help with the solubility of polysaccharides, which is achieved by (1) increasing the molecular affinity to water and (2) preventing the intermolecular association due to the electrostatic effects posed by the charged group.

Acidic polysaccharides are polysaccharides containing carboxyl groups (e.g., pectin) and/or sulfuric ester groups (e.g., carrageenan). As shown in **Figure 2**, pectin is a polysaccharide containing the majority of α-(1–4)-linked D-galacturonic acid unit in the backbone and a small percentage of branching portion (rhamnogalacturonan I and II). The acidic group may be free (or as a simple salt with sodium, potassium, calcium, or ammonium) or naturally esterified with methanol. Therefore, most of the natural occurring pectin is readily soluble in water due

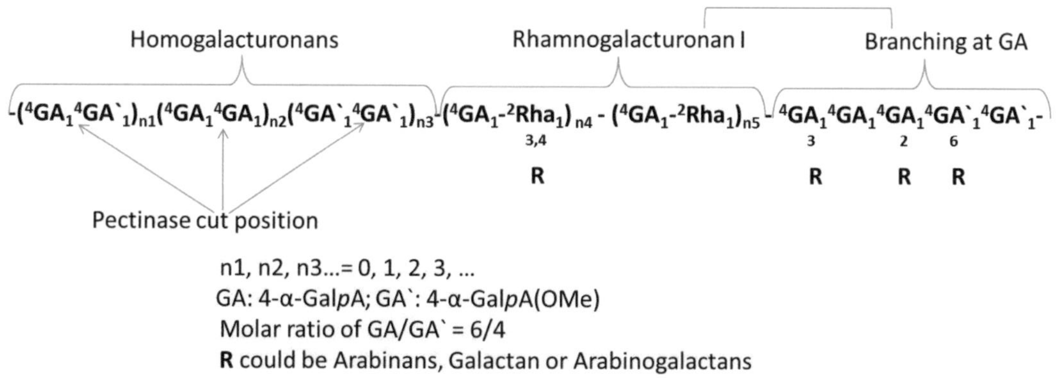

Figure 2. Pectic polysaccharides from American ginseng. Adpated from Guo, et al. [9].

to the charged group, although high in molecular weight. It also should be noticed that adding salt or reducing pH value could shield the charged effect, which leads to gelation under some circumstances. For example, high methyl ester pectin gel at pH 3.4 (rapid set pectin) or 3.2 (slow set pectin) with the presence of sucrose. Low methyl ester pectin can react with calcium ions to form gel, even under relative high pH environment. Therefore, when dissolving the pectin into water, it is essential to avoid the gelling condition; similar to other hydrocolloids, the dissolution usually needs high shearing mixing [8].

Carrageenans are a family of linear sulfated polysaccharides that are extracted from red edible seaweeds with the repeating unit of sulfate esters of 3-β-D-Gal*P* and 3,6-anhydro-α-D-Gal*P*. There are three types of commercial carrageenans including κ-carrageenan, λ- carrageenan, and ι-carrageenan, which showed different structural features regarding the number and position of sulfate groups in the repeating unit as shown in **Figure 3**. The higher levels of ester sulfate in λ carrageenan (three sulfate groups per repeating unit) contribute to gel inhibition and result in good water solubility. With the decrease of sulfate groups in the repeating unit, carrageenan start to form soft gel (ι-carrageenan, two sulfate groups per repeating unit) and rigid gel (κ-carrageenan, one sulfate group per repeating unit).

As one typical positively charged polysaccharide (**Figure 4**), chitosan is derived from the deacetylation of chitin. The positively charged groups come from protonation of its free amino groups, which is the key to its water solubility. Chitosan is insoluble in neutral and basic environments due to the lacking of a positive charge. However, in acidic environments, protonation of the amino groups increases the degree of water solubility. Following this property, chitosan has been widely used for drug delivery, e.g., the transport of insulin [11].

3.3. Branching polysaccharides vs. linear polysaccharides

The linear polysaccharides with highly regular conformation that can form crystalline or partial crystalline structures are mostly insoluble in water, while branching structure could increase the solubility for two reasons: (1) the branching structure could weaken the intramolecular interaction due to the steric effects, which prevent the intermolecular association, and

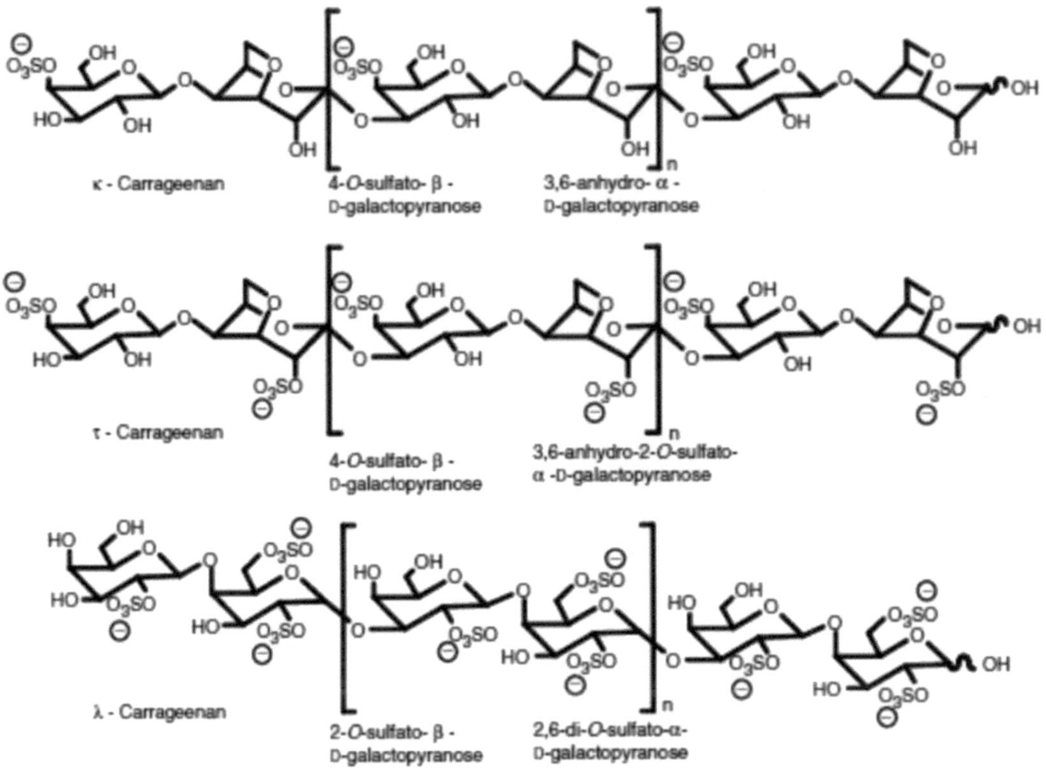

Figure 3. Comparison of structural features of κ-, λ-, and ι-carrageenan. Adopted from Cui et al. [10].

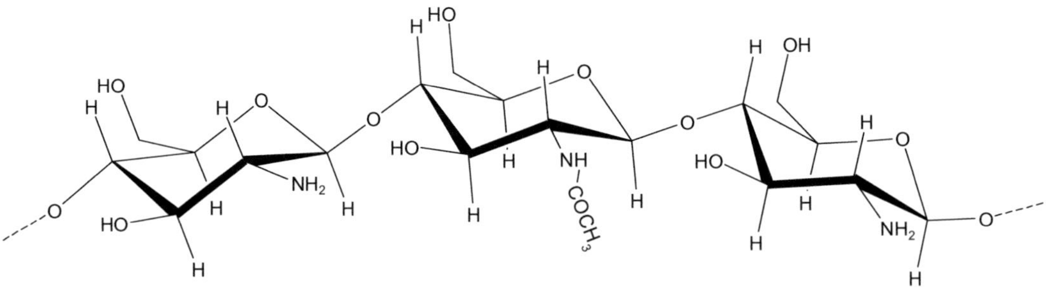

Figure 4. Structural feature of chitosan.

(2) the highly branched structure could also decrease the excluded volume when compared to polysaccharides with same molecular weight, which potentially increases the critical concentration and therefore improves the water solubility.

Taking cellulose as an example, cellulose with a regular (1 → 4)-β-D-glucan chain is essentially insoluble in aqueous medium. However, it can be modified by decreasing the Mw and introducing either charged (sodium carboxymethyl cellulose (CMC)) or branching groups (methyl cellulose (MC), hydroxylpropyl cellulose (HPC), hydroxylpropyl methyl cellulose (HPMC)) to increase the solubility (**Figure 5**). MC, HPMC, and HPC all can dissolve in cold water and are insoluble in hot water. MC and HPMC form gel when the temperature increases (52°C for MC and 63–80°C for HPMC), while HPC cannot form gel when heating, instead, it precipitates when temperature is above 45°C. Unlike MC and HPMC, CMC is well dissolved in both cold and hot water [12].

Starch contains both amylose and amylopectin. Amylopectin exhibits better solubility than amylose due to the highly branched structure, although the latter has relative low molecular weight (amylose, 10^5; amylopectin, 10^7–10^9). According to the structure and solubility difference, amylose and amylopectin can be separated from each other in starch granules according to the following procedure: firstly, starch granules are completely dispersed in hot water or aqueous dimethyl sulfoxide; amylose then can be precipitated by the addition of butanol (as a crystalline complex due to the linear structure) after cooling. Afterward, amylopectin can be recovered from the supernatant by lyophilization [13].

Guar gum and locust bean gum both belong to the galactomannan family (**Figure 6**), while the degree of branching for guar gum (galactose to mannose 2:1) is higher than locust bean gum (galactose to mannose about 4:1), which could easily prevent strong cohesion of the main backbones of different neighboring molecules, so that no extensive crystalline regions of guar gum can be formed, while locust bean gum is easy to form gel due to the naked region of the molecules, which favors the formation of junction zone [14].

Xylans of all higher plants possess (1–4) linked D-xyl*P* residues as the backbone, substituted by various degrees with sugar units including arabinose, xylose, and glucuronic acid (4-O-methyl). (1–4) linked D-xyl*P* as a linear chain showed the least solubility as it can be regarded as cellulose

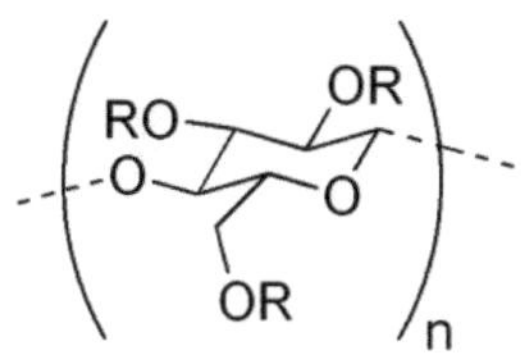

Figure 5. Schematic chart for cellulose derivatives. R=H or CH_3 for methyl cellulose; R=H or CH_2COOH for CMC; R=H or CH_3 or $CH_2CH(OH)CH_3$ for HPMC; and R=H or $CH_2CH(OH)CH_3$ for HPC.

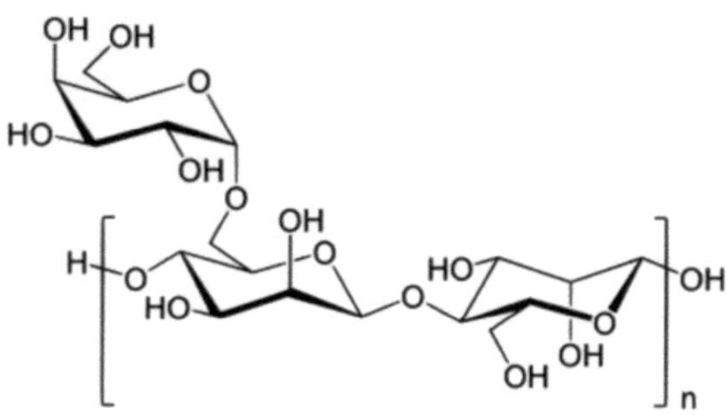

Figure 6. Schematic chart for galactomannan structure.

with hydroxymethyl group replaced with H. However, with the increase of degree of substitution such as arabinose (arabinoxylan) (as shown in **Figure 7**), its solubility dramatically increased [15].

Gum arabic has a highly branched structure (**Figure 8**). Although with a relatively high molecular weight, gum arabic showed excellent water solubility (up to 30% at room temperature). It has been reported that even 30% gum arabic solutions have a lower viscosity than 1% xanthan gum and sodium carboxyl methylcellulose at low shear rates. Gum arabic has been commercially used as emulsifiers due to the covalent bond with protein, in which protein functioned as hydrophobic group attached to the oil droplet and keeps the whole emulsion system stable. Similar to gum arabic, gum ghatti (**Figure 9**) also contains highly branched structure and demonstrated excellent water solubility (up to 20%). According to methylation analysis [16], the

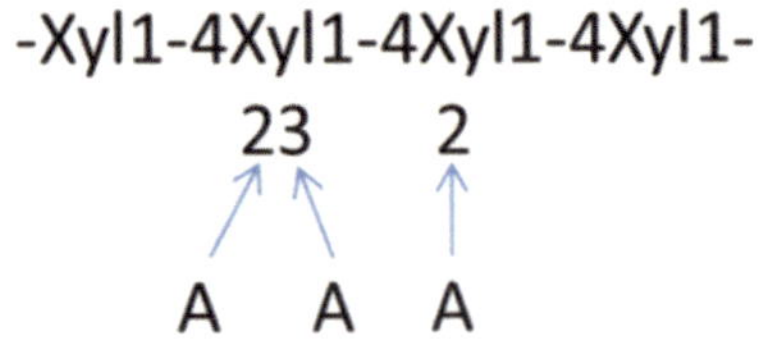

Figure 7. Schematic chart of arabinoxylan. A: T-Ara*P*.

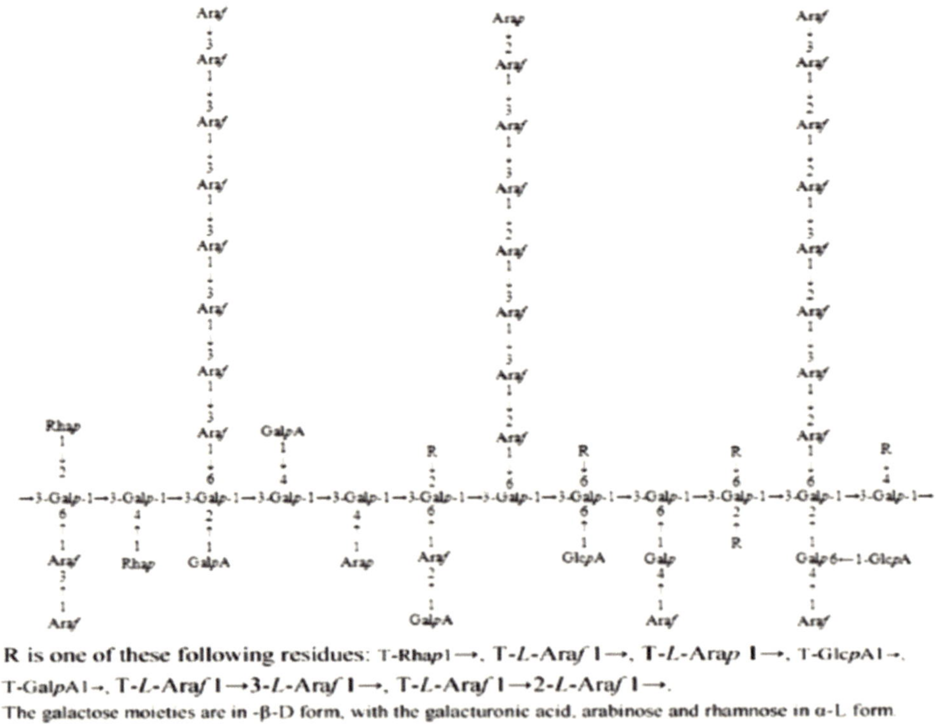

Figure 8. Proposed structure of gum arabic (Acacia senegal), adopted from Nie et al. [17].

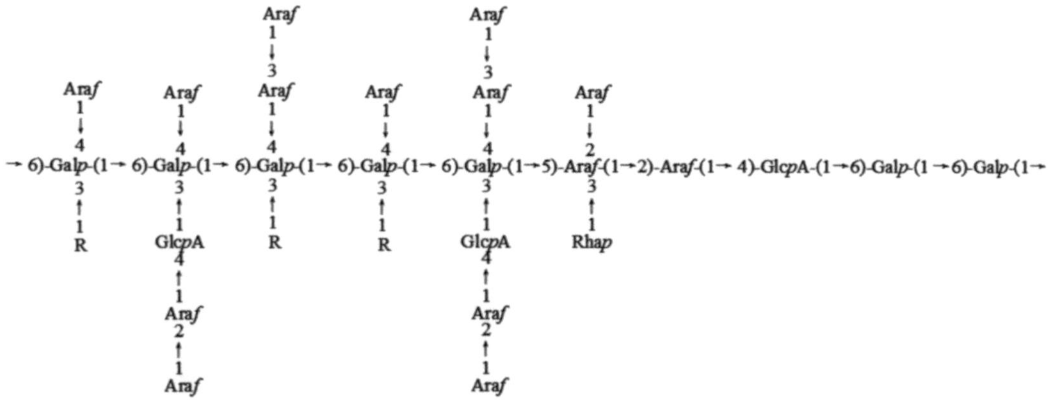

Figure 9. Proposed structure of gum ghatti, adopted from Kang et al. [18].

most branched sugar residues in gum ghatti were 3,4,6-Gal*p* (14.20%), 3,6-Gal*p* (5.10%), 2,3,6-Gal*p* (4.87%), 2,3,5-Ara*f* (4.09%), and 2,3-Man*p* (2.34%), respectively. The good solubility of gum ghatti could also be attributed to the 1–6 linked glycosidic bonds, which will be discussed in the later session.

3.4. Hydrophobic groups

Similar to branching effects, the presence of some hydrophobic groups, e.g., O-Ac and O-Me, could also affect the solubility of the polysaccharides.

O-acetyl substituents are present on many cell wall polymers including various hemicelluloses (**Figure 10**), the pectic polysaccharides and the polyphenol lignin, which have been previously summarized [19].

It has been reported that the totally acetylated xylan (DA 2.0) only dissolves in nonpolar solvents like chloroform or polar aprotic solvents like dimethyl sulfoxide, whereas the weakly acetylated xylan (DA ~0.5) is totally soluble in water, and the non-acetylated xylan (DA 0) is only partially soluble in hot water, due to spontaneous intramolecular hydrogen bonding [20].

Similar conclusion has also been drawn from acetylated glucomannan. One example is polysaccharides from *Dendrobium officinale* (traditional Chinese herbs), which belong to glucomannan family but highly acetylated, as shown in **Figure 11**. This polysaccharide is readily dissolved in water. The solubility, however, is significantly decreased if the acetyl group was either removed through alkaline treatment or increased through acetylation reaction. It has also been reported that the immunomodulating effect of this polysaccharide was also affected by the acetyl group [21].

3.5. Linkage pattern and conformation of sugar residues

The conformation can be simply regarded as the way that polymer chains align themselves in solution to adopt an orientation with lower energies. Two general types of conformation

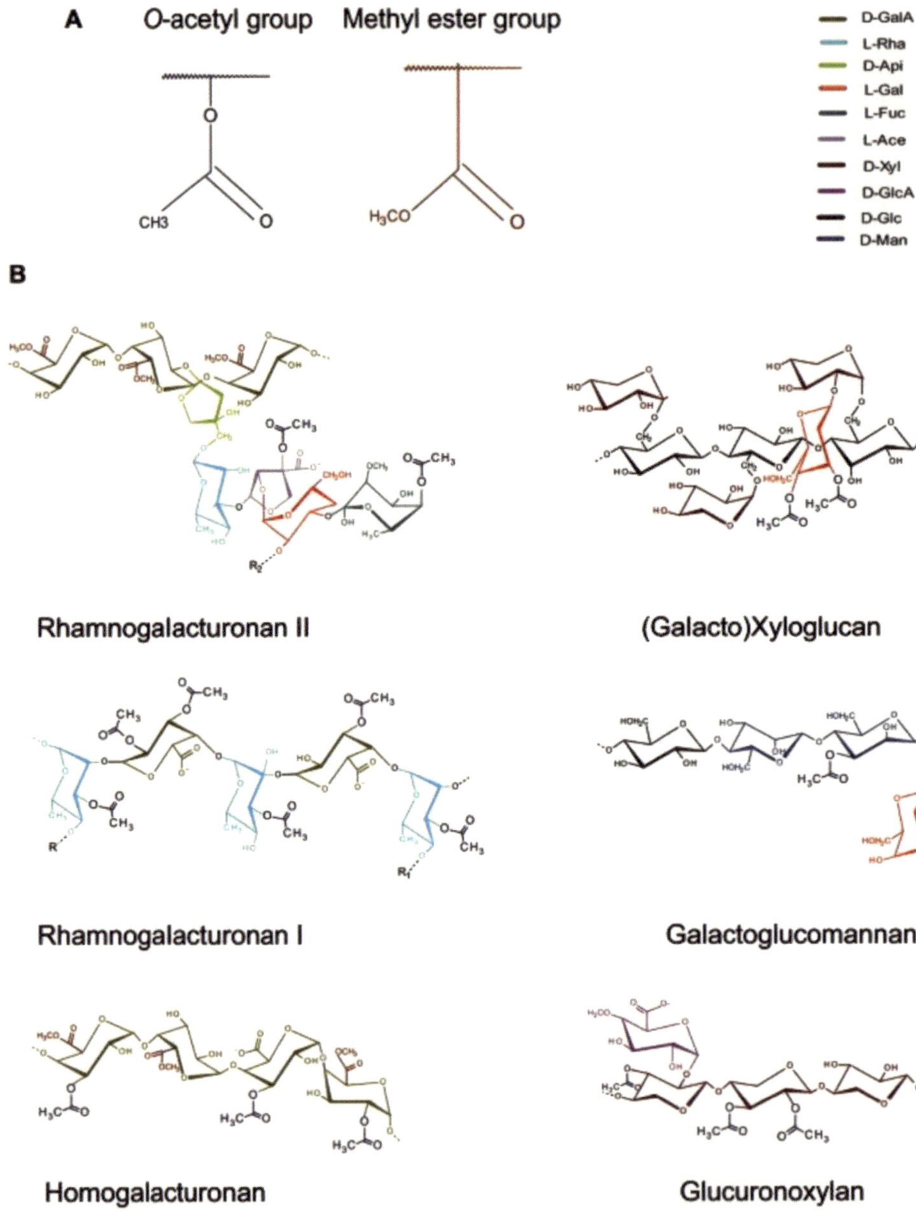

Figure 10. *O*-acetylation of cell wall polysaccharides. (A) Generic representation of *O*-acetyl group as found at different -OH positions in many cell wall polysaccharides. Note the structural similarity between *O*-acetyl- and methyl ester groups that decorate carboxylic acid residues in polygalacturonic acid. (B) Occurrence of *O*-acetyl groups in cell wall matrix polysaccharides. Adopted from Pawar et al. (2003) [20].

for polysaccharides can be simply divided—ordered conformation and disordered conformation—which is decided by the regularity of the molecular structure. In aqueous solution, most of non-starch polysaccharides with heterogeneous structure demonstrated disordered

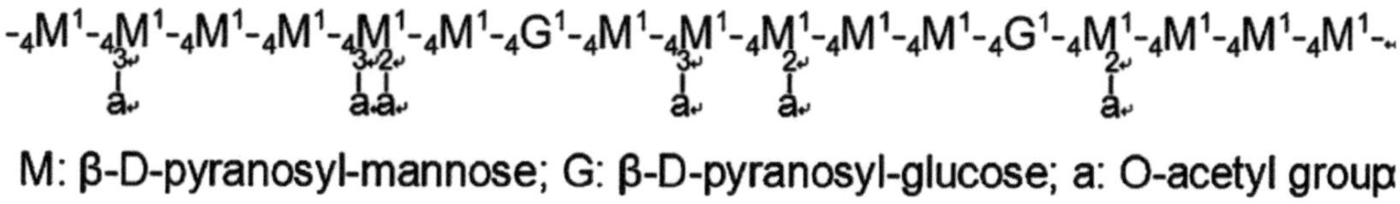

Figure 11. Proposed structure of polysaccharides from *Dendrobium officinale*. Adopted from Xing et al. [22].

conformation, including random coil, rigid, and spherical conformation. High-performance size exclusion chromatography (HPSEC) could be used to study the conformational properties of polysaccharides in aqueous solution. Combined refractive index (RI) light scattering detectors (LALS and RALS) with an online viscometer, the relationship of M_w and the intrinsic viscosity [η], and R_g, R_h and R_g/R_h (ϱ) can be obtained. The double logarithmic plot of the molecular weight vs. intrinsic viscosity could be well described using the Mark-Houwink equation (8):

$$[\eta] = kM_v^{\alpha} \tag{8}$$

where k and α could reflect the corresponding conformation of these polysaccharides. It is well known that the exponent α normally lies in the range of 0.5 to 0.8 for linear random coil polysaccharides and increased with increasing chain stiffness. Polysaccharides with α below 0.5 and above 0.8 are generally regarded as spherical and rigid chain conformation. α is also affected by the interaction between polysaccharides and solvent. Low α value favors poor solvent, and high α indicates a good solvent. The logarithmic plot of the molecular weight versus intrinsic viscosity of heteropolysaccharide isolated from seeds of *Artemisia sphaerocephala* Krasch is shown in **Figure 12**, from which a random coil conformation of this polysaccharide was determined. However, the curve in **Figure 12** is not exact linear, a slightly decreased slope was observed with the increase of Mw, which was attributed to the increased percentage of side chains in the high Mw fraction by the authors [3].

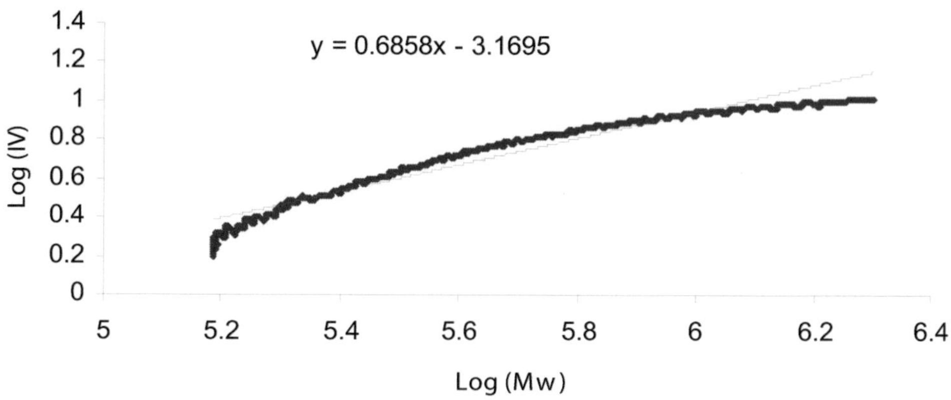

Figure 12. Logarithmic plot of the molecular weight vs. intrinsic viscosity of weight heteropolysaccharide isolated from seeds of *Artemisia sphaerocephala* Krasch. Adopted from Guo et al. (2013) [3].

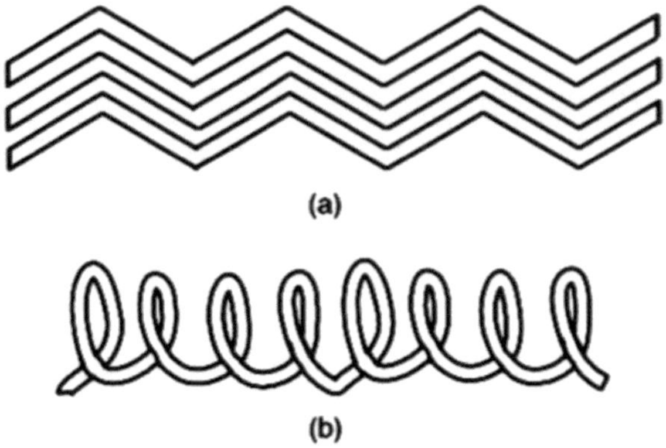

Figure 13. Linkage patterns and conformation types of polysaccharides: (a) ribbon-like and (b) hollow helix. Adopted from Cui and Wang [23].

If the values of the torsion angles are fixed by cooperative interactions between residues, such as in solid or gel states, an ordered conformation can be adopted. Two general ordered conformations are ribbon-like and helix conformations (**Figure 13**). Polysaccharide with ribbon-like conformation is most easily aligned and closely packed through numerous hydrogen bonds and van der waals forces. The resultant compact structures essentially prevent solvent penetration and retain insolubility in water. Polysaccharides such as cellulose, xylan, and mannan, which contained the zig-zag type linkages, all belong to this type. Another ordered conformation is hollow helix conformation, in which the bonds adopt a U-turn form, as in 3-β-D-Glc*P* (curdlan) and 4-α-D-Glc*P* (amylose from starch). Compared to ribbon-like conformation, the hollow helix conformation showed a relative better solubility, but still not comparable with disordered conformation (random coil), e.g. amylose adopts the helix conformation and is only dissolvable in hot water [23] (**Figure 13**).

Polysaccharides with the same monosaccharides but different linkage patterns (solubility) also showed varied solubility. Compared to cellulose, the solubility of cereal β-glucan is increased with the insertion of $1 \rightarrow 3$ linkage bond into the $1 \rightarrow 4$ glyosidic bond. Generally speaking, any structures that contain especially flexible units such as $(1 \rightarrow 6)$ linkages lead to easier solubility because of a large favorable entropy of the solution. For example, by introducing single β-D-Gal*P* substituents $(1 \rightarrow 6)$ linked to mannan backbone, the resulting galactomannans are fairly soluble in water [24]. Pulluan is another very typical example, which is also known as α-1,4- and α-1,6-glucan, and prone to dissolve in water.

4. Discussions

It can be seen that polysaccharides have very complex structural features regarding monosaccharide composition, glycosidic bonds (linkage patterns), the degree of branching, molecular weight distribution, and the presence of special groups. These structural features highly affected the solubility of polysaccharides, in either positive or negative way. All in all, any structure that hinders the intermolecular association usually leads to a higher solubility, such as branching structure, charged group (carboxylate group, sulfate, or phosphate groups); on the opposite, structural characters that promote the intermolecular association result in the poor solubility, such as linear chain, large molecular weight, and other regular structural characters. It should be noted that these structural features are not isolated but intercorrelated with each other to decide the solubility of any specific polysaccharide. For example, the excellent solubility of gum arabic is not only contributed by its highly branched character but also affected by the charged group as well as the 1–6 linkage bonds. The good solubility of amylopectin is contributed by both highly branched structure and $1 \rightarrow 6$ linkage bonds. Therefore, it is wise to consider the relationships of solubility with overall polysaccharide structure features. Some examples of polysaccharides with different solubility groups have been previously summarized [7], while some modification has been made (**Table 1**).

Understanding the relationships between solubility and structural features of polysaccharides could better facilitate the food and nonfood applications. The solubility of some originally insoluble or poor soluble polysaccharides such as cellulose and starch can be increased through basic structural modification. Take starch for example; starch can be physically or chemically modified to improve water solubility. The physical modification methods involve the treatment of native starch granules under different temperature/moisture combinations, pressure, shear, and irradiation, while chemical modification involved with acid treatment, cross-linking, oxidation, and substitution (esterification and etherification) to either decrease the starch molecular weight, increasing the substitution or increase the charged groups.

Solubility groups	Polysaccharide	Monomer	Special/charged group	Branching situation	Conformation
Poor solubility	Cellulose	4-β-D-Glc*P*	No	Linear	Ribbon
	Xylan	4-β-D-Xly*P*	No	Linear	Ribbon
	Mannan	4-β-D-Man*P*	No	Linear	Ribbon
Intermedia	Cereal β-glucan	4-β-D-Glc*P*,3-β-D-Glc*P*	No	Linear	Random coil
	Konjac glucomannan	4-β-D-Glc*P*,4-β-D-Man*P*	Acetyl group	Linear	Random coil
	Locus bean gum (galactomannan)	4-β-D-Man*P*,T-α-D-Gal*P*	No	Less and short branched	Random coil
Good solubility	Gum arabic/gum ghatti	3,4,6-, 3,6- Gal*p*,-Ara*f*, and so on	Glucuronic acid	Highly branched	Spherical
	Pullulan	4-α-D-Glc*P*,6- α-D-Glc*P*	No	Branched	Random coil
	Amylopectin	4-α-D-Glc*P*,6- α-D-Glc*P*	No	Highly branched	Spherical

Table 1. Examples of polysaccharides with different solubility.

Author details

Mark Q. Guo[1], Xinzhong Hu[2]*, Changlu Wang[1] and Lianzhong Ai[3]

*Address all correspondence to: hxinzhong@126.com

1 School of Food Engineering and Biological Technology, Tianjin University of Science & Technology, Tianjin, China

2 College of Food Engineering and Nutritional Science, Shaanxi Normal University, Xi'an, China

3 School of Medical Instrument and Food Engineering, University of Shanghai for Science and Technology, Shanghai, China

References

[1] Miller-Chou BA, Koenig JL. A review of polymer dissolution. Progress in Polymer Science. 2003;**28**(8):1223-1270

[2] Podzimek S. Polymers, in Light Scattering, Size Exclusion Chromatography and Asymmetric Flow Field Flow Fractionation. Hoboken, NJ: Wiley-VCH; 2011. p. 1-36

[3] Guo Q, et al. Conformational properties of high molecular weight heteropolysaccharide isolated from seeds of *Artemisia sphaerocephala* Krasch. Food Hydrocolloids. 2013;**32**(1):155-161

[4] Ding HH, et al. Xyloglucans from flaxseed kernel cell wall: Structural and conformational characterisation. Carbohydrate Polymers. 2016;**151**:538-545

[5] Li W, et al. Studies of aggregation behaviours of cereal β-glucans in dilute aqueous solutions by light scattering: Part I. Structure effects. Food Hydrocolloids. 2011;**25**(2):189-195

[6] Kang J, et al. New studies on gum ghatti (*Anogeissus latifolia*) part 5: The conformational properties of gum ghatti. Food Hydrocolloids. 2015;**43**:25-30

[7] Whistler RL. Solubility of polysaccharides and their behavior in solution. In: Isbell HS, editor. Carbohydrate in Solution. Washington, DC: American Chemical Society Press; 1973. p. 242-255

[8] Endreß HU, Christensen SH. 12 – Pectins. In: Phillips GO, Williams PA, editors. Handbook of Hydrocolloids (Second Edition). Boca Raton: CRC; 2009. p. 274-297

[9] Guo Q, Cui SW, Kang J, Ding H, Wang Q, Wang C. Non-starch polysaccharides from American ginseng: Physicochemical investigation and structural characterisation. Food Hydrocolloids. 2015:**44**;320-327

[10] Cui S, Izydorczyk M, Wang Q. Polysaccharide gums: Structures, functional properties, and applications. In: Cui S, editor. Food Carbohydrates : Chemistry, Physical Properties, and Applications. Boca Raton: CRC; 2005. p. 263-307

[11] Agnihotri SA, Mallikarjuna NN, Aminabhavi TM. Recent advances on chitosan-based micro- and nanoparticles in drug delivery. Journal of Controlled Release. 2004;**100**(1):5-28

[12] Murray JCF. 25 – Cellulosics. In: Phillips GO, Williams PA, editors. Handbook of Hydrocolloids (Second Edition). Boca Raton: CRC; 2009. p. 710-723

[13] Liu Q. Understanding starches and their role in foods. In: Cui S, editor. Food Carbohydrates: Chemistry, Physical Properties, and Applications. Boca Raton: CRC; 2005. p. 309-356

[14] Wielinga WC. 10 - Galactomannans. In: Phillips GO, Williams PA, editors. Handbook of Hydrocolloids (Second Edition). Boca Raton: CRC; 2009. p. 228-251

[15] Izydorczyk MS. 23 – Arabinoxylans. In: Phillips GO, Williams PA, editors. Handbook of Hydrocolloids (Second Edition). Boca Raton: CRC; 2009. p. 653-692

[16] Kang J, et al. Understanding the structure-emulsification relationship of gum ghatti - a review of recent advances. Food Hydrocolloids. 2014;**42**:187-195

[17] Nie SP, et al. The core carbohydrate structure of Acacia Seyal Var. Seyal (gum arabic). Food Hydrocolloids. 2013;**32**(2):221-227

[18] Kang J, et al. New studies on gum ghatti (*Anogeissus latifolia*) part II. Structure characterization of an arabinogalactan from the gum by 1D, 2D NMR spectroscopy and methylation analysis. Food Hydrocolloids. 2011;**25**(8):1991-1998

[19] Gille S, Pauly M. O-acetylation of plant cell wall polysaccharides. Frontiers in Plant Science. 2012;**3**:12

[20] Pawar, et al. Acetylation of woody lignocellulose: Significance and regulation. Frontiers in Plant Science. 2013;**4**:118

[21] Xing X, et al. A review of isolation process, structural characteristics, and bioactivities of water-soluble polysaccharides from *Dendrobium* plants. Bioactive Carbohydrates and Dietary Fibre. 2013;**1**(2):131–147

[22] Xing X, et al. Study on *Dendrobium officinale* O-acetyl-glucomannan (Dendronan((R))): Part II. Fine structures of O-acetylated residues. Carbohydrate Polymers 2015;**117**:422–433.

[23] Cui S, Wang Q. Understanding the conformation of polysaccharides. In: Cui S, editor. Food Carbohydrates: Chemistry, Physical Properties, and Applications. Boca Raton: CRC; 2005. p. 219-261

[24] Cui S, Wang Q. Understanding the physical properties of food polysaccharides. In: Cui S, editor. Food Carbohydrates: Chemistry, Physical Properties, and Applications. CRC Press; 2005. p. 161-218

Chapter 3

Osmotic Properties of Polysaccharides Solutions

Kruk Joanna, Pancerz Michał and Ptaszek Anna

Additional information is available at the end of the chapter

http://dx.doi.org/10.5772/intechopen.69864

Abstract

Osmotic properties of polysaccharides' solutions and associated biopolymer-solvent and biopolymer-biopolymer type interaction are very important from a technological point of view. The knowledge of osmotic properties of these systems provides the basis to appropriate use of polysaccharides having comply with the relevant technology functions, impart the appropriate texture and forming the sensory properties of the final product. Furthermore, an important issue is the effect of time on the osmotic properties of polysaccharides', because with time, the aforementioned effects may change. Membrane osmometry is one of the methods used in the studies of synthetic polymers to determine their average molecular mass and the degree of interaction between a polymer and a solvent. This method is successfully applied in the case of biopolymers that include polysaccharides. The existence of the osmotic pressure, formed by diffusion of solvent molecules through a semi-permeable membrane, is the basis of this method. Analysis and interpretation of osmometric research results is based on the van't Hoff equation dependency of the concentration. The second virial coefficient obtained based on this relation allows characterisation of biopolymer-solvent interactions, and thus biopolymer tendency to solvation. The third virial coefficient provides information on mutual interactions between the biopolymer molecules, as well as its tendency to aggregate.

Keywords: osmotic pressure, virial coefficient, overlap concentration

1. Introduction

Osmotic properties are part of wider group of colligative phenomena and concern the liquid-vapour equilibrium in multi-component systems. This group of colligative properties includes depression of the freezing point (cryoscopy), boiling point elevation (ebullioscopy) and osmotic pressure in general. The osmotic pressure can be associated with water activity of

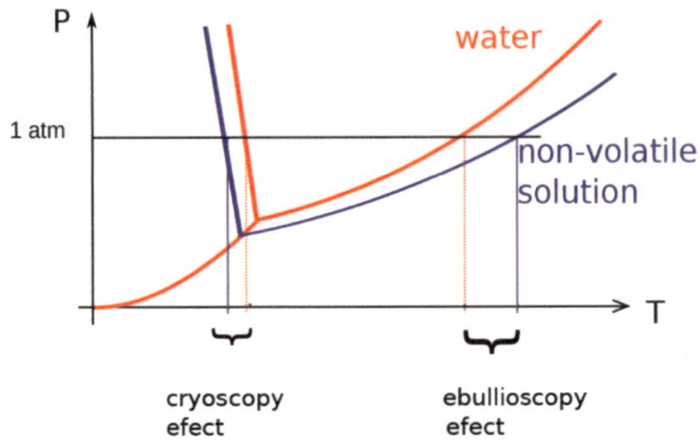

Figure 1. Phase equilibrium for pure water and solution.

various types of products. The essence of the discussed phenomena is related to the changes of pressure of the saturated vapour (**Figure 1**), which causes the dissolution of the non-volatile substance.

If the non-volatile substance is a low-molecular-weight chemical compound, the changes in the vapour pressure can be explained by common phenomena, such as association or solvation, which are a result of interactions between molecules. The difference of the vapour pressure would in this case be directly proportional to the molecular mass of a dissolved substance. The osmotic measurements can therefore be used to determine the molar mass, or for multi-molecular substances with significant polydispersion, to determine the average osmotic molecular mass. The osmosis process takes place between the solution and a clear solvent, or between solutions of different concentrations, provided they are separated by a membrane, which is permeable only to the solvents molecules (**Figure 2**). The solvent moves through the membrane from the solution with lower concentration of the dissolved substance (or from the area of clear solvent) towards the solution with higher concentration. From the point of view of the molecules present in the solution, it is a phenomenon opposite to diffusion. From the solvents perspective, it is a natural intent to balance the chemical potentials, which results in the dilution of the solution with higher concentration. If the osmotic equilibrium takes place in

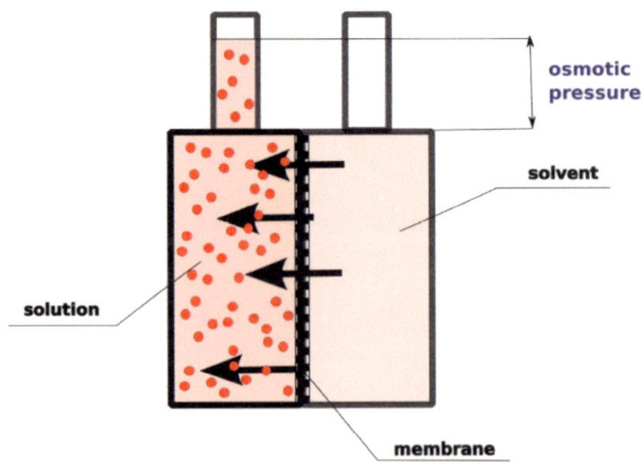

Figure 2. Osmosis mechanism.

a solution-clear solvent system, the dilution of the solution becomes so significant that the dissolved substances presence becomes suppressed and as a result, the pressure of solvents vapours over the solution becomes equal to the vapour pressure over the clear solvent. Because of that, the osmotic pressure is considered as one of the colligative processes.

2. Measurements of osmotic properties

The measurements of osmotic pressure can be carried out using two types of osmometers: membrane and vapour (**Figure 3**).

A membrane osmometer consists of two chambers divided by a membrane with specific pore sizes, which allow the solvent to move. One of the chambers is filled with pure solvent, while the other with the studied solution and the difference in the hydrostatic pressure between the two chambers is measured. These devices can be used to determine the pressure up to about 0.1 mmH_2O, which in practice means measuring polymer and biopolymer solutions of up to 2000 kDa. The lower measuring range depends on the membrane's permeability. In the case of these devices, the membrane itself is the source of the main issues. The membrane's permeability depends on its structure: whether it has a system of pores (inorganic materials) or is a molecular sieve (organic materials). Permeability is usually expressed by the lowest molecular mass of a substance that the membrane allows through (*cut-off* value). For biopolymer solutions characterised by high polydispersion, there is a possibility of diffusion of fractions with lower molecular mass through the membrane. If it is close to the solvent's molecular mass, the obtained value of osmotic pressure is not affected. The solvent's flow through the membrane can cause a so-called *balloon effect*, manifested by bulging of the membrane and distortion of its surface, which can affect its properties. Another issue is caused by the viscosity effect of the solution, which for biopolymers forming structural fluids, even at their low concentrations,

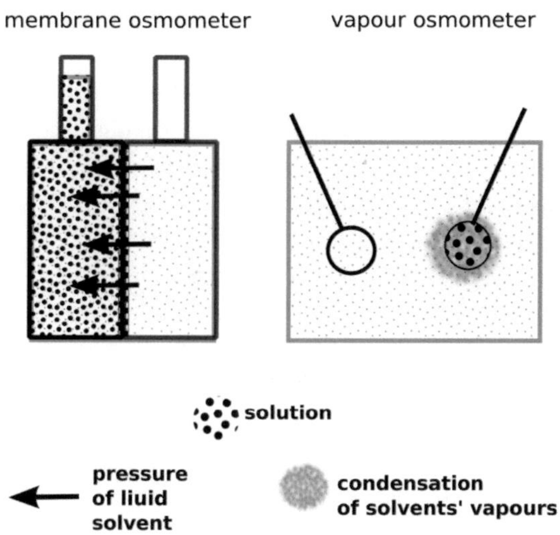

Figure 3. Types of osmometers.

can negatively affect the measurements due to the long time required for the establishment of osmotic equilibrium. Reaching the equilibrium is directly related to the solvent molecules diffusion throughout the solution. The vapour osmometer operates on a different premise: it measures the heating effect of the condensation of solvent's vapour on the drop of the tested solution, as well as a clear solvent. The osmometer consists of a chamber filled with the solvents vapours, in which two capillaries are placed (**Figure 3**). The capillaries are filled with the solvent and the tested solution, respectively, in such a way that drops form on their ends. In a state of thermodynamic equilibrium, the rate of solvent's evaporation and the rate of the condensation of its vapours would be the same. The whole process is carried out at a characteristic temperature—the boiling point (at a given pressure). On the other side, an intensive condensation of the solvent's vapours on the solution drop occurs, caused by the difference of the vapour pressure, which results in a local increase of temperature in the near vicinity of the drop (compared to the boiling temperature of pure solvent). It is that difference of temperatures, that is, the basis of the measurements in the vapour osmometer. Devices of this type can be used to carry out measurements of molecular masses in the range from 40 to 40 kDa. For these devices, one of the limitations is the viscous effect of the tested solution, which for biopolymers plays a significant role. Moreover, the measurement range of these devices makes them practically impossible to use to study multi-molecular substances.

3. Osmotic equilibrium equation

The test results obtained from the osmotic pressure measurements are analysed based on the virial equation, which is mainly associated with real gases. While it might seem surprising, the validity of this approach is based on the nature of interactions between molecules, which move chaotically and collide with each other. In the real gases, the collisions between molecules are not elastic, and as a consequence, they shape the characteristics of the gas phase, so that they diverge from the Clapeyron's law. For liquid solutions, a similar interpretation can be applied: the polymer/biopolymer molecules are incomparably larger than the solvent molecules (colloid solutions), and therefore, the solvent becomes the `background' for the interactions between macromolecules. For a macromolecule-solvent system, the nature of the interactions is more complex. Numerous studies on the structure of macromolecules in solutions indicate to two possible behaviour patterns, dependent on the interactions between the solvent and the polymer/biopolymer chains (**Figure 4A**). As a result of those interactions, the polymer chain can undergo either expansion or contraction (collapse). The scenario is decided by the affinity to the solvent. The chain's expansion is related to the absorption of the solvents molecules in between the chain's segments and can be caused by solvation, electrostatic (Couloumb) interactions, formation of a helical structure or the presence of a spatial hindrance in the case of branched polymers (**Figure 4A**). A contraction caused by low affinity to the solvent induces a collapse, which is often accompanied by the aggregation of the chains or creation of a rigid branched structure, which leads to phase separation (**Figure 4A**).

The quality of a solvent is examined in close relation to a specific polymer/biopolymer. A good solvent (**Figure 4B**) causes the expansion of the chain in the solution, which in the range of

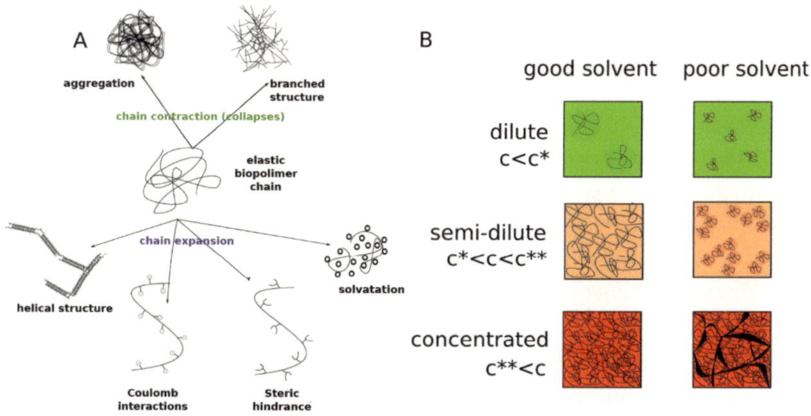

Figure 4. Swelling mechanism (A) and quality of solvent (B).

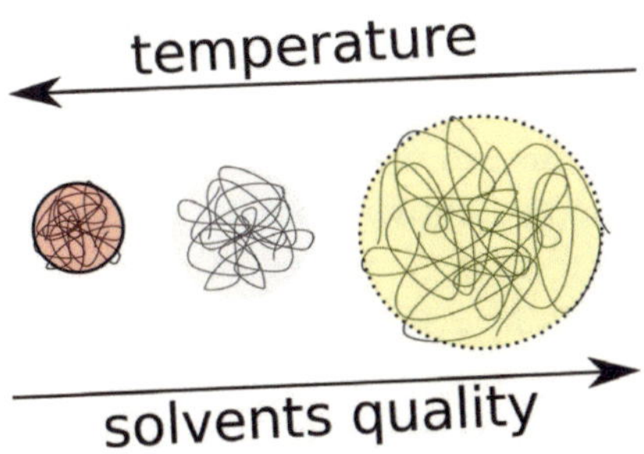

Figure 5. Swelling as a function of temperature and solvents quality.

concentrations $c < c^*$ (diluted solution) can undisturbedly interact with the solvent. Exceeding the critical concentration limit c^* (*overlap concentration*) causes a change in the interactions to those characteristic of a semi-dilute system. The value of c^* (*overlap concentration*) is dependent on the biopolymer's molecular mass and represents average concentration of individual cluster segments. In the case of the lower the molecular mass, the higher the overlap concentration c^* means a lower concentration of chains, which when exceeded, results in diffusion (good solvent, **Figure 4B**) or aggregation (poor solvent, **Figure 4B**) of the molecules. Crossing the second overlap concentration c^{**} results in a formation of a concentrated solution, whose structure is close to a system of entanglements (good solvent) or phase separation is observed, which is caused by incompatibility (poor solvent). Conformation of the chain in a solution is dependent on the balance between osmotic pressure in the chain's close proximity (solvent absorption) and its elasticity, either enabling expansion or limiting it. The discussed behaviour is strongly related to temperature and the quality of the solvent, which is represented in **Figure 5**.

All the outlined consequences of the solvent-macromolecule and macromolecule-macromolecule interactions mean that the osmotic state equation should, as extensively as possible, take into account the interactions shaping the properties of a tested solution. Because of that, the applied virial state equation takes the following form:

$$\frac{\pi}{c_W} = \frac{RT}{M_n}\left[1 + A_2(T)c_W + A_3(T)c_W^2 + \ldots\right] \tag{1}$$

Its form is analogous to the virial equation of real gas. From a mathematical point of view, it can be considered as an expansion of the function into a series, around a solution corresponding to ideal solution behaviour. In that equation, $A_2(T)$ and $A_3(T)$ correspond to the second and third virial coefficients, c_W to biopolymer concentration, and M_n to the average osmotic molecular mass of the diluted substance. A detailed interpretation of Eq. (1) enables the analysis of the interactions between the diluted substance and the solvent and the molecules of the diluted substance. In the simplest interpretation, a negative value of $A_2(T)$ indicates a low affinity between a polymer and a solvent and in consequence, a probability of aggregation/association of the chains, or even precipitation or recrystallisation. High positive values of that coefficient, in turn, indicate a full compatibility of the solvent and the macromolecule.

4. Starch polysaccharides and osmotic properties of its pastes

Starch, which is produced and stored in plants seeds and bulbs, is one of the most important plant polysaccharides. It is composed of two types of biopolymers: amylose (AM) and amylopectin (AP). Amylose is a linear fraction composed of α-D-glucose units, bonded to each other through $\alpha(1 \rightarrow 4)$ glycosidic bonds, and its molecular mass is in the range from 100 to 1000 kg/mol (10^5–10^6 Da). Amylopectin has a higher molecular mass: 10^3–10^5 kg/mol (10^6–10^8 Da) and is heavily branched with $\alpha(1 \rightarrow 6)$ bonds [1–3]. There are only few known systems in which starch is soluble. They include water, dimethyl sulfoxide (DMSO) and N,N-dimethylacetamide with the addition of LiCl [4]. From the food industry perspective, the most important starch solvent is water. Dissolving starch requires a complete destruction of the starch granule, which is composed of layers of crystalline structure and amorphic layers [1–3]. The effect is a solution comprising both linear and branched chains. Their conformations and behaviour are different, and both are related to the type of solvent. Amylose, which in water solutions adopts a double helix conformation, starts to aggregate immediately after leaving the granule (precipitation combined with the formation of a crystalline structure, retrogradation [5] and after exceeding, overlap concentration (c^*) undergoes the gellation process [6, 7]. Amylopectin's conformation can be compared to a cluster with side chains sticking out. As a result, a structure is formed in which protruding chains of neighbouring macromolecules intersect and form an amorphic network of entanglements (**Figure 4**). The results of studies by Prof. Burchard's team seem to confirm above described phenomenon. (**Table 1**). A comparison of dimensions of $R_{g;branched}$ amylopectins with the $R_{g;linear}$ amyloses of the same molecular mass M_w:

$$g = \left(\frac{R^2_{g;\,branched}}{R^2_{g;\,linear}}\right)_{|M_w} \tag{2}$$

leads to a conclusion that AP chains exhibit smaller dimensions ($g < 1$) than amylose. The water solutions of amylopectin are stable and the chain intersections occur after exceeding the

Starch	AM, %	M_w, kg/mol	R_g, nm	A_2, 10^6 mol mL g^{-2}	c^*, g/L	g
Amylose*	100	150–750	25	–	–	–
Potato	24	51,000	222	8.96	7.9	0.166
Corn	22	88,000	213	3.65	3.1	0.077
Waxy corn	0	76,900	234	4.20	3.1	0.130
Amylstarch	76	16,700	231	3.14	17.0	0.168

The data concerns water starch solutions tested with the light scattering method SLS 20°C [2, 3]. M_w—weight average molecular mass, R_g—gyriation radius, c^*—overlap concentration, A_2—second virial coefficient; all values were determined based on SLS measurements. *The results shown for amylose were carried out with chromatography methods.

Table 1. Molecular parameters of starch with varying amylose (AM) content.

overlap concentration c^* during a long period of time. If the concentration of AP in the solution is lower than c^*, then amylopectin clusters undergo a collapse [3] caused by the '*squeezing*' of water molecules.

For linear amylose, the interactions between its chains result in a formation of cluster structures (**Figure 4A**). Because of that, the molecular masses determined with the use of light scattering measurements (SLS) are significantly higher than those determined chromatographically for a single chain. The level of aggregation of AM chains depends on the initial concentration, at the start of dissolution. Dissolution of AM in water requires application of high temperatures of 135°C, as well as pressure. The authors of the Ref. [2] study carried out autoclaving at 10–15 bar. These distinct conformations of AM and AP raise a question about the nature of these chains' coexistence in water solutions: do amylose and amylopectin form separate structures and therefore are two separate components or whether they form a blend. Tests on solutions and pastes indicate the latter; however, phase separation is observed in those systems (**Figure 6**).

Starch solutions obtained with the use of DMSO do not exhibit incompatibility with the solvent [8]. This behaviour is a result of the fact that the chains do not exhibit a tendency to aggregate. The values of the average radii of gyration and M_w obtained from SLS tests (**Table 2**) and chromatography are comparable [9]. A very small average radius of gyration obtained for the wax potato starch indicates a low level of branching of that amylopectin [10]. The test results presented in the Bello-Perez et al. [10] study indicate that both corn and potato amylopectins adopt spherical or globular conformation in DMSO. Amylose in DMSO adopts an

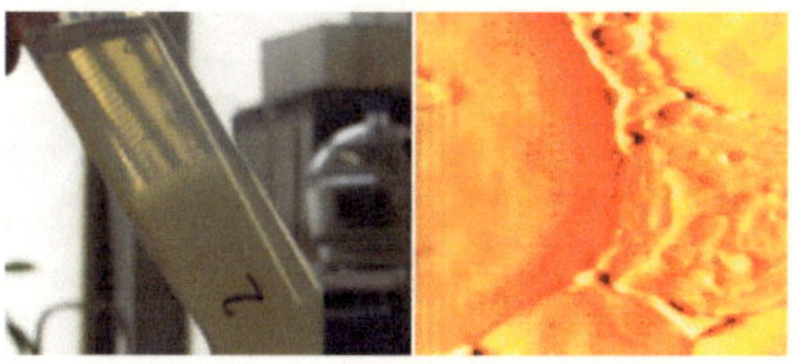

Figure 6. Incompatibility phenomenon in starch paste, AFM image of regular maize starch paste (right, own study, GONOTEC Qscope, USA, non-contact mode).

Starch	AM, %	M_w, kg/mol	R_g, nm	A_2, 10^3 mol mL g^{-2}
Amylose[1]	100	108–235	24–31	0.87–0.94
Potato[1]	24	26,000	127	0.0148
Waxy potato[2]	0	4000	15	–
Waxy corn[2]	0	53,000	242	–

The data concerns starch solutions in DMSO tested with light scattering method SLS at 20°C [7][1] and [8][2].

Table 2. Molecular parameters of starch with varying amylose (AM) content.

elastic chain conformation [8]. This finding is supported by the high values of the second virial coefficient (**Table 2**).

The mixture of DMSO and water can be used as a special solvent. These solutions exhibit a strongly non-ideal behaviour, due to the interactions between molecules. The methyl groups of DMSO may induce cooperative ordering in the system by hydrophobic hydration effects. The oxygen atom of a DMSO molecule can interact with water through H-bonding [10] with the continuum percolation transition in aqueous DMSO solutions with the percolation threshold of 12–15 wt.% of DMSO. A number of studies have been carried out over the years to understand the conformational properties of linear amylose [10–14] (**Table 3**) and branched amylopectin [14–17], in water and DMSO mixtures. The authors determined that the temperature and time of the dissolution have substantial influence on the weight average molecular mass, radius of gyration and the dispersion of the polysaccharide chains in the solution. With an increasing addition of water, the interactions between amylose and DMSO were reduced, leading to the conformational transition of AM from tight helical via loose helical to disordered coil [11]. The AP's solubility is limited by the presence of water in the solvent [17]. Increasing water content not only limits the AP's solubility but also causes an aggregation of its chains in

	Solution H_2O/DMSO	M_w, kg/mol	R_g, nm	A_2, 10^6 mol mL g^{-2}	α^2	Reference
	100/900	765	37.5	272	2.192	
AM	200/800	660	38.8	276	1.952	[16]
	500/500	555	34.0	123	1.304	
	700/300	506	26.3	55.6	1.189	
	100/900	151	84	–		[13]
	100/900	15,300	99.8			
	300/700	57,500	182.3			[19]
AP	500/500	192,700	182.8			
	100/900	171,000	238			[18]
	100/900	150,000	238	0.055		[13]

α^2—expansion coefficient.

Table 3. Molecular parameters of amylose and amylopectin in binary solvents H_2O/DMSO tested with light scattering method SLS at 25°C.

the solution. In the case of starch solutions in binary solvents, the phenomenon of coil overlap occurs.

In the case of utilising pure DMSO or DMSO/water mixtures, it is possible to achieve a decidedly higher solubility of starch and its derivatives than for pure water—up to 50% (w/w) starch solutions in pure DMSO [20] and in binary solvents ([21], the below photo from own studies).

For amylose solutions (**Figure 7**), a change in character of the π/c dependency can be observed: from ascending in the range of lowest concentration to descending for $c > 0.05$ g = 100 mL. The value of the second virial coefficient A_2 for the amylose solutions, determined based on the osmotic equation of state (EoS), is positive (**Table 4**). The third virial coefficient is negative, which indicates a possibility of aggregation or even recrystallisation of amylose in water solutions [6]. Additionally, this phenomenon is supported by the average osmotic molecular mass, whose value was estimated based on Eq. (1) to be 5500 kDa. The π/c relationships observed for AP and potato starch exhibit particularly interesting courses. In particular, it can be noted that the values of osmotic pressure are negative for certain concentrations of the polysaccharide. The negative values of π are related with the direction of flow of water in the measurement chamber of the device. The measurement of π consists of determining the pressure applied by the solvent, which flows to the measurement chamber through the membrane. When it comes to a negative signal, an opposite phenomenon is observed: a '*squeezing*' of the solvent from the solution and its flow through the membrane into the chamber containing clear solvent. A phenomenon of water squeezing and collapse of the biopolymer coils was described in Ref. [2]. The critical concentration (*overlap concentration, c**), at which this phenomenon ceases, at

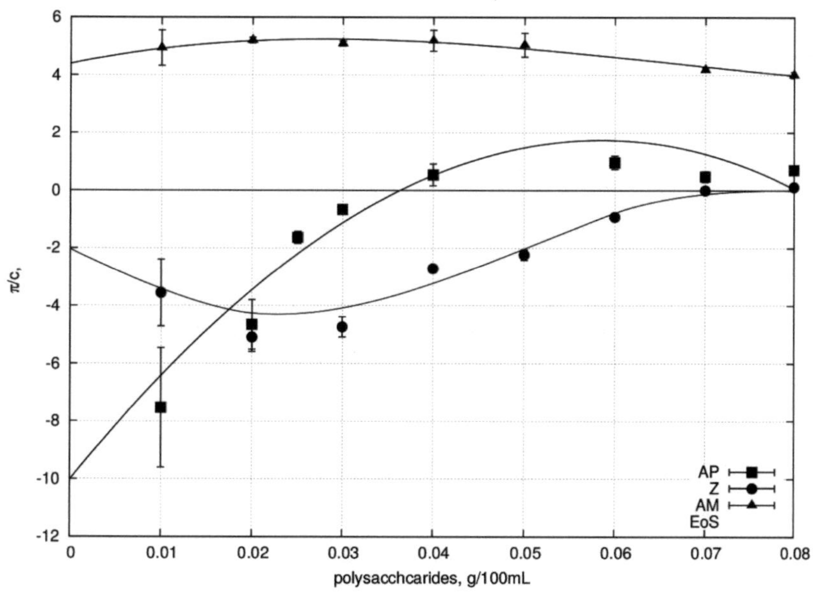

Figure 7. Correlation between reduced osmotic pressure π and concentration of water starch solutions (measurements carried out at 30°C using membrane osmometer GONOTEC Osmomat 090).

	Day	AM	AP	Starch native	Ac	Ph	E1404	M_w
30°C	1	$2.05{\cdot}10^{-6}$	$1.34{\cdot}10^{-5}$	$-1.09{\cdot}10^{-5}$	$4.17{\cdot}10^{-6}$	$1.06{\cdot}10^{-6}$	$-1.84{\cdot}10^{-6}$	$7.47{\cdot}10^{-7}$
	2		$4.42{\cdot}10^{-5}$	$4.17{\cdot}10^{-5}$				
	8		$4.17{\cdot}10^{-5}$	$4.48{\cdot}10^{-5}$				
40°C	1		$4.25{\cdot}10^{-5}$	$1.87{\cdot}10^{-5}$				
	2		$3.32{\cdot}10^{-5}$	$2.62{\cdot}10^{-5}$				
	8		$2.68{\cdot}10^{-5}$	$3.07{\cdot}10^{-5}$				

Table 4. Value of the second osmotic virial coefficient A_2 mol mL g^{-2} or native starches (amylose AM, amylopectin AP) and modified starches (acetylated Ac, starch phosphate Ph, starches oxidised with E1404 and oxidised with microwave radiation M_w).

25°C was: for AP 3.1 mg/mL (c^* = 0.31 g/100 mL) and for potato starch 7.9 mg/mL (c^* = 0.79 g/100 mL) [2, 3]. The concentrations observed on the charts, for which the π/c curve exceeds zero value, were 0.038 g/100 mL for AP and 0.074 g/100 mL for potato starch (at 30°C) and are therefore qualitatively consistent with the results of studies by Burchard and co-workers. This indicates that a concentration increase fosters the interactions between polysaccharide chains. Because of negative values of π, it was not possible to estimate the values of average osmotic molecular masses. Values of second and third virial coefficient were determined for amylopectin and potato starch pastes. At 30°C, the A_2 for AP measured $1.34{\cdot}10^{-5}$ mol mL g^{-2} and $A_3 < 0$. These results point to a high affinity of amylopectin to water (positive value of the second coefficient), with the ability to form local bundles of amylopectin clusters (negative third coefficient). These observations are consistent with literature results [20]. For the potato starch, the $A_2 = -1.09{\cdot}10^{-5}$ mol mL g^{-2} and $A_3 > 0$. These values indicate a gellation process caused by the presence of amylose.

4.1. The effect of time on osmotic properties of pastes

Storage of the pastes in room temperature causes changes in the interactions between their components (**Figure 8**). The results of osmotic pressure measurements at 30°C carried out 24 h after the first measurements (*day* 2) are similar to the behaviour of a fresh paste, however in the range of lower concentrations, a much lower osmotic pressure is observed. This probably indicates an increase of collapse of amylopectin chains. The overlap concentration for which the osmotic pressure becomes greater than zero has not changed. For *day-8* measurements, the osmotic pressure is not greater than zero in the analysed range of concentrations. The observed behaviours are the characteristic of dissolved solutions of amylopectin clusters, because they occur in the range of concentrations lower than overlap concentration, which for the reference temperature 25°C is the literature stated at 0.31 g/100 mL [2, 3]. However, positive values of A_2 (**Table 4**) indicate that water is a good solvent for amylopectin and are consistent with literature data. The measurements of osmotic pressure were also carried out at 40°C. The results are shown in **Figure 8**.

No qualitative changes in the course of the π/c curve were noted for the *fresh* solution compared to the relationship plotted at 30°C. Overlap concentration c^* increases to the value of

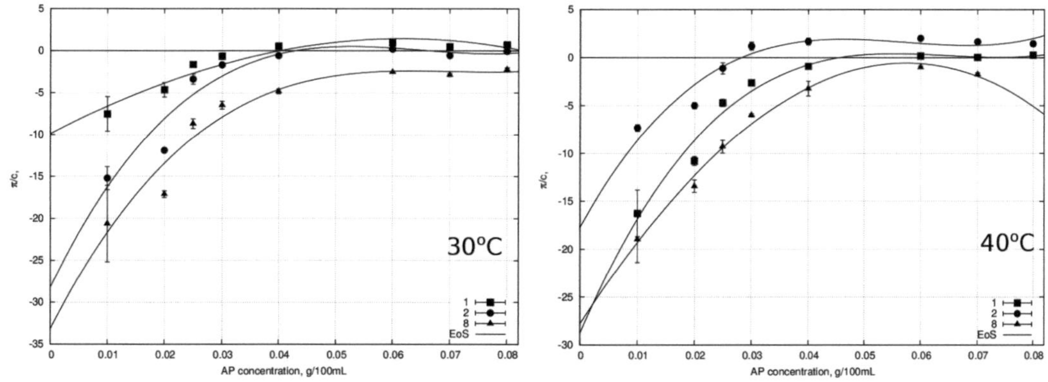

Figure 8. Correlation between reduced osmotic pressure π and water solution of amylopectin (AP) in the function of storage time of the solution (measurements carried out at 30 and 40°C using membrane osmometer GONOTEC Osmomat 090).

0.045 g/100 mL and the value of the second virial coefficient to $4.25 \cdot 10^{-5}$ mol mL g^{-2}. On the second day, the value of overlap concentration c^* decreases abruptly to 0.03 g/100 mL, which indicates an increase of *critical dimensions* of amylopectin clusters. The value of the second coefficient is still positive (**Table 4**), while $A_3 < 0$. After 8 days of storage, even an increase in measurement temperature does not result in significant qualitative or quantitative changes in the course of π/c in the function of concentration in comparison to the results of tests carried out at 30°C. Only the value of A_2 decreases slightly (**Table 4**), and the third virial coefficient is negative.

The measurements of π changes were also carried out for the potato starch pastes (**Figure 9**). On the second day of storage, a dramatic change in the interactions of solution's components was observed. After exceeding $c^* = 0.026$ g/100 mL, the osmotic pressure becomes positive. The course of the π/c curve points to two observations. Firstly, the critical radius of gyration of starch increases significantly, which is a result of a massive change in the overlap concentration in relation to the value obtained for *day 1*. This is related to the joint retrogradation of amylose

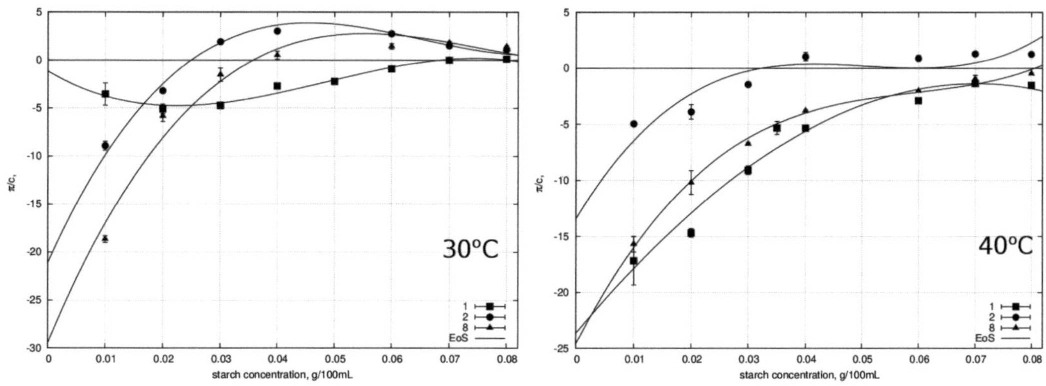

Figure 9. Correlation between the reduced osmotic pressure π and the concentration of water potato starch solution in the function of storage time (measurements carried out at 30 and 40°C, using a membrane GONOTEC Osmomat 090).

and amylopectin. Most probably, the presence of AP as a spatial hindrance accelerates the AM recrystallisation process. Secondly, an occurrence of behaviour characteristic of gelling systems is clearly observed because an increase of π/c value is noted followed by its decrease, similarly to the solutions of pure AM (**Figure 7**). The values of the second and third virial coefficients (**Table 4**) become positive. On *day 8* of storage, the value of the second virial coefficient is positive, while $A_3 < 0$. The value of c^* increases to 0.032 g/100 mL, which indicates a reduction of critical dimensions of chains present in the solution. Similar measurements carried out at 40°C indicate to structural changes at higher temperatures. The value of the second virial coefficient, determined based on measurement results carried out on the first day, is positive. However, the values of π are negative, and this is related to the *squeezing* of water from the clusters. In the discussed range of concentrations, the value of c^* cannot be determined. This could indicate a certain relaxation of the structure with an increase of distance between clusters. Properties of the mixture after 24 h of storage undergo changes. The values of π become positive for $c = 0.035$ g/100 mL, which indicates an increase of critical dimensions of the aggregates. On day 8, the occurrence of solvent *squeezing* is predominant (negative values of π) due to starch retrogradation.

Chemical modification of starch causes changes in the interactions between polysaccharide chains and water (**Figure 10**). For phosphorylated potato starch, the reduced osmotic pressure has negative values. This behaviour is different to the native potato starch, as for the tested range of concentrations (0.050–0.075 g/100 mL), the positive values of osmotic pressure were not observed. Based on that, it is possible to speculate that phosphorylation has caused the change to the overlap concentration, at which values of π would adopt positive values. This

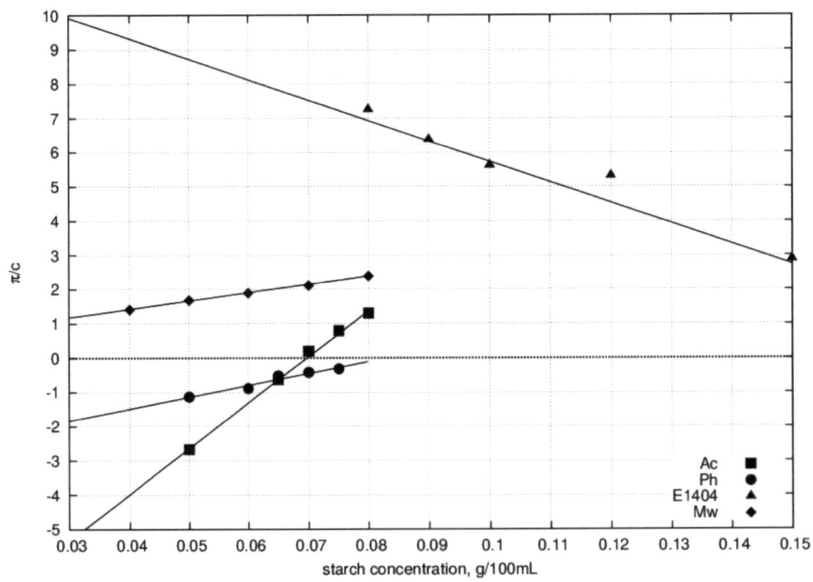

Figure 10. Correlation between reduced osmotic pressure π and concentration of water potato starch solutions: acetylated (Ac), phosphorylated (Ph), oxidised (E1404) and subjected to microwave radiation (M_w), in the function of concentration (measurements at 40°C).

implies, that in comparison to native potato starch, the chains of phosphorylated starch undergo contractions, also at higher concentrations of the solution. The second virial coefficient calculated for the obtained test data is positive, which indicates a change in the interactions between the solvent and polysaccharide. An increase in solvation of the biopolymer chains occurs.

Acetylation of starch, similarly to phosphorylation, causes a change to the concentration at which the values of osmotic pressure are positive. This results in a change of osmotic properties, which is in accordance to the results presented by Żmudziński et al. [22]. The values of reduced osmotic pressure are negative for concentrations in the range from 0.050 to 0.065 g/100 mL and positive from 0.070 to 0.080 g/100 mL. The correlation obtained for the tested systems is ascending for the whole range of concentrations. Positive values indicate an increase of water absorbability of the tested solutions. The second virial coefficient calculated for these systems has a positive value. An increase of the affinity of the solvent molecules to the polymer chains occurs, which means that water is a good solvent for acetylated starch.

For solutions of starch oxidised in a microwave radiation field, the obtained π/c values are positive in the whole range of concentrations. The obtained π/c correlation is ascending. This behaviour indicates that this starch modification in these particular conditions causes an increase in water absorption by the potato starch chains, which can be related to an easier access of solvent to the starch granules. The determined virial coefficient is positive, which allows us to deduce an improvement of permeability and solvation of starch chains due to the microwave radiation.

The values of the reduced osmotic pressure obtained for the oxidised starch (E1404) are the highest of all tested systems; however, the obtained correlation is descending. It is worth noting that the concentration range included higher concentrations than for the other systems. A common concentration for the tested starches is 0.080 g/100 mL, for which the value of π/c is approximately seven times higher than for the acetylated starch and approximately four times higher than for the starch treated with microwaves. The obtained results indicate to the absorption of water by the starch chains. The second virial coefficient is negative but higher than for native starch, which indicates approximately 10 times more dissolution capability than native potato starch.

5. Non-starch polysaccharides

The non-starch polysaccharides group includes gums of various plant origin: locust bean gum, konjac, guar gum and carrageenans or bacterial origin polysaccharide: xanthan and chemical modified cellulose: carboxylmethylcellulose.

Locust bean gum (carob gum) is obtained by the milling of endosperm of the seeds from the carob tree pods (*Ceratonia siliqua L.*, belonging to *Fabaceae* family), vegetation characteristic for the Mediterranean regions. The endosperm in the seeds takes approximately 42–46%, the remaining parts are bran husk 30–33% and germ 23–25% [23]. The extraction of carob gum can be carried out by two methods. The first comprises thermo-mechanical or chemical separation of the bran husk, and then separation of the germ from the endosperm and finally subjecting the endosperm to milling, sieving and sorting. The second method uses a combination of extraction and purification [24, 25]. Whole brans are subjected to milling and then extracted using water and precipitation with the use of alcohol (ethanol or isopropanol) [26–28]. The flour is a non-ionic biopolymer belonging to galactomannans; therefore, the main chain of this polysaccharide is formed by mannose units joined by β-(1,4)-glycosidic linkages, to which single units of D-galactopyranose are attached by the α-(1,6) linkages. In the carob gum, the side branches are not positioned symmetrically; therefore, unsubstituted β-D-mannopyranose units can occur [29, 30]. The ratio of mannose to glucose ranges between 3.1 and 3.9 [31–33] and depends on the variety, source and most of all on the applied extraction method [33]. The galactose content of approximately 20% is lower in comparison to other galactomannans, such as guar gum and tara gum [28, 34]. The ratio of mannose to galactose (M/G) has impact on this polysaccharides solubility in water [31, 35], the higher ratio of (M/G) results in better thickening properties [35]. Side chains in the form of attached mannose units at the C-6 carbon position contain arabinose residues [31]. Galactomannans exhibit a polymer structure of '*random coils*'; therefore, high temperature and energetic mixing are required for their complete dissolution in water (to achieve best water binding) [34, 37]. The locust bean gum was the first to be used not only in food industry but also in the production of textiles, paper, pharmaceuticals and cosmetics [36, 38].

Guar gum is a non-ionic polysaccharide obtained from the endosperm of guar seeds (*Cyamopsis tetragonoloba*, from the family *Leguminosae*). The extraction methods are similar to those used for extraction of the locust bean gum and comprise separation of shells, milling and extraction [34, 39], sometimes enzymatic processing or extrusion is also utilised. The main chain of this polysaccharide is formed by D-mannose units linked by β-(1,4)-glycosidic linkages and also contains single units of D-galactose, attached to the main chain by the α-(1,6)-glycosidic linkages. The ratio of mannose to glucose is approximately between 1.5:1 and 1.8:1 [29, 40] or approximately 2:1 [36]. The ratio of mannose to glucose together with the molar mass of this polysaccharide depends on the variety of the seed from which it is extracted and affects its solubility, thermal stability and rheological properties [38, 41]. Guar gum is widely used in the food, cosmetic and mining industries [43–45].

Konjac gum is obtained from the tubers of the *Amorphophallus konjac* plant, belonging to the *Araceae* family and farmed in Asia. The tubers of the konjac plant grow and enlarge throughout the plants lifetime. Among the products obtained from konjac tubers, depending on the purification degree, are: konjac flour, konjac gum and purified konjac glucomannan. Two-year-old tubers are used for the production of konjac flour, as after 3–5 years they bloom and are used as seed material. The content of konjac glucomannan in the bulbs ranges from 8 to 10%, and the remaining components are starch, proteins, fats and mineral components [43, 46]. The main component of konjac gum is the konjac glucomannan, a neutral heteropolysaccharide formed by D-mannose and D-glucose, joined by β-(1,4)-glycosidic linkages. Molar ratio of mannose to glucose is 1.6:1 [47–50]. The main chain is slightly branched by glucose units in the C-6 carbon position [51, 52]. The branching can also occur in the C-3 carbon position, both for mannose and glucose [42]. The degree of branching of the polysaccharide chain is approximately 8%. Moreover, native glucomannan exhibits a slight degree of acetylation (DS) from 0.05 to 0.1 (1 in 12 or 18 sugar units) [51, 53, 54]. The chain acetylation affects its solubility. Konjac gum is soluble in water and, due to high possibilities of water binding, forms solutions of high viscosity [50, 55]. Similarly to other hydrocolloids, such as xanthan gum or carrageenan, it forms gels [56, 57]. It is widely used in food, pharmaceutical and cosmetic industries [51, 58].

Gum arabic is a natural plant sap excretion harvested from acacia trees (*Acacia Senegal* and *Acacia seyal*) [59–60] and is obtained by cutting holes into the bark. The sticky sap dries on the trees, forming hard lumps, which are collected and sorted and then grounded. Another production method comprises the dissolution of gum arabic in water at low temperatures, so that its denaturation would not occur, and then, the obtained solution is subjected to decantation or filtration. Afterwards, it is subjected to pasteurisation and in the last stage, the mixture is subjected to drying [53, 61]. The structure of the gum arabic is rather complex. The main chain of this polysaccharide is formed by (1,3) and (1,6) β-D galactopyranose units, connected with (1,6) β-D-glucopyranose units of uronic acid. The side chains can contain units of α-L-ramnopyranose, β-D-glucuronic acid, β-D-galactopyranose units and α-L-arabinofuranose units [59, 62–64]. The main characteristic of gum arabic is its covalent bond with the protein chain [59]. It is known that a protein molecule rich in hydroxyproline (Hyp), serine (Ser) and proline (Pro) constitutes the core, to which polysaccharide units connect via Ara-Hyp (Ara—arabinose) linkages. The polysaccharide chains coil into globular units and '*decorate*' the protein chain in such a way as to achieve an overall spheroidal shape [59, 66]. Gum arabic is highly soluble in water (up to 50% w/v), and its solutions have relatively low viscosity in comparison to other plant gums. This is a result of the highly branched molecular structure and relatively low molecular mass of this polymer [54]. The protein part of the gum arabic is responsible for the surface activity and has impact on the foam forming and emulsifying properties of this polymer [63, 67, 68]. Currently, it is used as a food additive, a component in pharmaceutical industry and is also utilised for technical purposes [64, 69].

Xanthan gum is an extra-cellular polysaccharide produced by the *Xanthomonas campestris* bacteria on the cell wall surface during a complex enzymatic process [70, 71]. On the industrial scale, xanthan gum is produced with the use of pure bacteria cultures in a submerged aerobic fermentation process. The bacteria cultivation happens in an adequately aerated environment containing glucose, nitrogen sources and other trace elements. In order to ensure an adequate efficiency, the fermentation process is divided in several stages. After the last stage of fermentation, the obtained bullion is pasteurised and the xanthan gum is extracted by precipitation with isopropyl alcohol. The final stages of production include drying and grinding [54]. The basic structure of xanthan gum includes cellulose skeleton formed of β-(1,4) D-glucose units, substituted alternatively by glucose with trisaccharide chain units at C-3 carbon position. This chain is formed by two mannose units, separated by (D)-glucuronic acid. The acid is attached to the external (D)-mannose unit by a β-(1,4) linkage, and to the internal one by the β-(1,2) glycosidic bond, which in turn is bound to the glucose of the main chain by a β-(3,1)glycosidic linkage [72, 73]. Approximately half of the end mannose units at C-6 carbon have an attached purine group, while the internal units at C-6 carbon position have acetyl groups attached. Xanthan gum is therefore a pentamer, and each molecule contains approximately 7000 of them [65]. The chain of xanthan gum may also have a structure of single or triple helix [66]. It is soluble in cold water forming solutions of high apparent viscosity (i.e., 1% solution 1.2–1.6 Pa·s). Solutions of xanthan gum are pseudo-plastic liquids, this is a result of the molecules ability to form aggregates by hydrogen bonds and polymer entanglements. Thanks to that, a pseudogel structure is formed, allowing practical uses of xanthan gum solutions [54].

Carboxymethylcellulose (CMC) is a half-synthetic anionic polysaccharide obtained by chemical modification of cellulose. During the process, a partial substitution by carboxymethyl group occurs of the second, third and sixth hydroxyl groups in cellulose. The linear chains of CMC are formed of glucose units joined by β-(1,4)glycosidic linkage. Average substitution CMC level is defined as an average number of carboxymethyl groups repeating unit. This parameter is defined in the range 0.4–1.5. This polymer is usually available in the form of sodium salt, and the product soluble in water is characterised by the substitution level above 0.5. Molar mass of this polymer is varied, for example, it can be in the range $2.5{\cdot}10^{-5}$–$7.0{\cdot}10^{-5}$ g mol^{-1}. CMC is widely used in the food, cosmetic, paper, textile and drilling industries [76, 77].

Carrageenans are a group of polysaccharides, in which 15 types can be distinguished, differing between each other based on structure. These polysaccharides are obtained by alkaline extraction of red edible seaweeds. The ι-, κ-, λ-carrageenan are mainly used in practical application. The main native polysaccharide chain of these three fractions is formed by a repeated poly-(1,3)-[4-sulphate-β-D-galactopyranose-(1,4)-3,6-anhydro-2-sulphate-α-D-galactopyranose]. κ-Carrageenan poly-(1,3)-[4-sulphate-β-D-galactopyranose-(1,4)-3,6-anhydro-α-D-galactopyranose]. λ-Carrageenan poly-(1,3)-[2-sulphate-β-D-galactopyranose-(1,4)-2,6-anhydro-α-D-galactopyra nose. κ- and ι-Carrageenans, as opposed to the lambda fraction, form thermo-reversible gels. Above the melting temperature, the chains of this polymer occur in the form of coils, and during the lowering of temperature, the coils change the conformation to double helices and form larger aggregates and thus form a spatial gel network structure [78, 79].

5.1. Osmotic properties of nonstarch polysaccharides' solutions

The tests on the osmotic properties of the polysaccharides water solutions were carried out using a membrane osmometer at two temperatures: 30 and 40°C. For the solutions of locust bean gum (LbG), the relation of the reduced osmotic pressure (π/c) to the concentration (c) at both temperatures was descending in the whole range of tested concentrations. It was observed, that after exceeding c^* (1.5), the values of π/c undergo little changes (for the three highest concentrations). In that range of concentrations, higher values of π/c were noted at temperature of 40°C, which could be explained by the fact that increased temperature causes lower aggregation of biopolymer chains than the 30°C temperature. This interpretation of the phenomenon is also influenced by the interactions of biopolymer-biopolymer type. Below c^*, two concentrations can be distinguished, at which a change in the behaviour of the tested solutions occurs. Namely, for the concentration of 0.1 g/100 mL, the reduced osmotic pressure assumes a higher value at 40°C. The reversal takes place at the concentration of 1.5 g/100 mL, and a return to higher values of π/c at 40°C was observed for the concentration of 0.4 g/100 mL>c^*. The values of the second virial coefficient determined for both temperatures are negative (**Table 5**). This indicates to a low affinity of the solvent to the polysaccharide chains. An increase of measurement temperature from 30 to 40°C causes a small change of the value of this

	M_w, g mol^{-1}	c^*, g dL^{-1}	Reference	30°C	40°C	M_w^{osm}, g mol^{-1}
KG	10^5–10^6	0.08	[87, 89]	$-2.52 \cdot 10^{-6}$	$-8.14 \cdot 10^{-7}$	$0.24 \cdot 10^5$
GG	$9.1 \cdot 10^5$	0.13				
	$7.3 \cdot 10^5$	0.28		$-2.43 \cdot 10^{-7}$	$-5.40 \cdot 10^{-7}$	$1.6 \cdot 10^5$
	$4.0 \cdot 10^5$	0.45	[70]			
	$2.7 \cdot 10^5$	0.45				
LbG	$0.697 \cdot 10^6$–$0.94 \cdot 10^6$	0.4	[71]	$-2.02 \cdot 10^{-7}$	$-2.20 \cdot 10^{-7}$	$1.4 \cdot 10^5$
	$1.94 \cdot 10^6$–$2.29 \cdot 10^6$	0.33	[72]			
XG		0.03	[72, 84]	$-1.36 \cdot 10^{-4}$	$-8.02 \cdot 10^{-5}$	
CMC	$3.5 \cdot 10^4$–$1.7 \cdot 10^6$	–	[68]	$-9.84 \cdot 10^{-5}$	$-1.58 \cdot 10^{-4}$	$2.57 \cdot 10^6$
AG	$5 \cdot 10^5$–$6.5 \cdot 10^5$	–	[74]	$2.80 \cdot 10^{-7}$		$1.07 \cdot 10^6$
CA	$6.6 \cdot 10^3$–$1.2 \cdot 10^6$	–	[75]	$-8.41 \cdot 10^{-6}$		$0.08 \cdot 10^5$

Table 5. Average molecular masses, values of the first overlap concentration and values of the second osmotic virial coefficient A_2 mol mL g^{-2} for:

coefficient, which suggests that in that range of temperatures, only a small increase of solubility of LbG chains in water takes place (**Figure 11**).

In the case of guar gum solutions (GG), the measurements were done in the same range of concentrations as for LbG, both <c^* as well as >c^*. The value (c^*) is confirmed by the study by Baines and Morris [76]. The values of reduced osmotic pressure exhibit a decreasing tendency with an increase of polysaccharide concentration. Initially (concentration range 0.1–0.3 g/100 mL), the reduced osmotic pressure assumes higher values at 30°C temperature. The reversal of this behaviour occurs only at the 0.4 g/100 mL concentration. In the final range of concentrations (0.45–0.5 g/100 mL), the highest values of π/c were noted at 30°C temperature, which indicates a different behaviour than in the case of LbG. Exceeding c^* results in an increase of the degree of chain aggregation and an increase of influence of biopolymer-biopolymer interactions, the effects of which are more visible at 40°C temperature. The second virial coefficient assumes

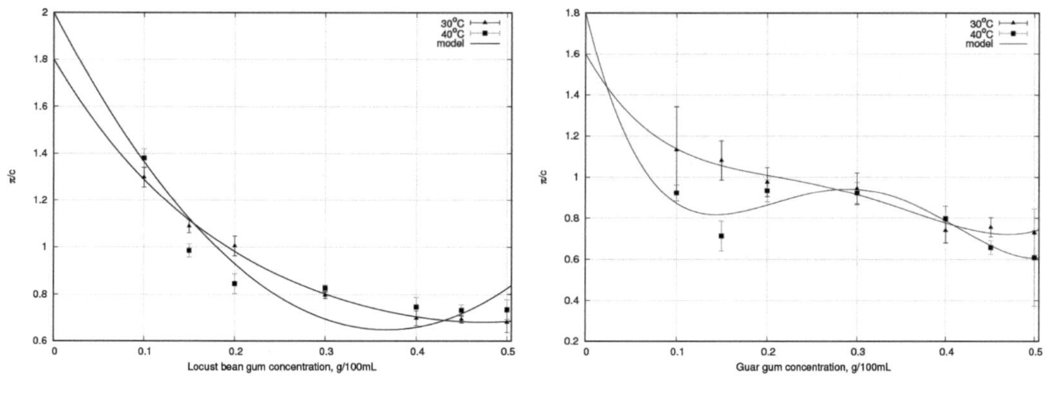

Figure 11. Correlation of reduced osmotic pressure π and the concentration of water solutions of locust bean gum and guar gum in the function of concentration (measurements at 30 and 40°C).

negative values for both temperatures [80–83]. At 40°C, a two times lower value was noted for this polysaccharide, than at 30°C, which can be interpreted as a decrease of the tendency of GG chains to solvation. An increase in temperature, therefore, causes an increase in solubility. Exceeding c^* (in our case 0.3 g/100 mL) causes the values of π/c at 40°C temperature to be similar to those obtained at 30°C. At concentration of 0.4 g/100 mL, the value of π/c is visibly higher at 40°C than at 30°C. This could be explained by the fact that for a certain range of concentrations, after exceeding c^*, an increase of temperature causes the infiltration of solvent particles into the chains. However, above the concentration of 0.4 g/100 mL, due to an increasing force of biopolymer-biopolymer interactions, a decrease of the π/c value is observed. Both for GG and LbG, we are looking at a '*random coil*' chain structure, which impedes the penetration of the solvent into the polysaccharide chain. Low values of π/c (0.6–1.4 $mmH_2O/g\ dL^{-1}$) for both gums indicate low water absorption of these polysaccharides in the studied range of concentrations (**Figure 12**).

Xanthan gum solutions (XG) were tested in the range of concentrations below c^* (**Table 5**). The relation of the reduced osmotic pressure and the concentration at 40°C temperature, as for the above described cases, is descending (**Figure 13**). At 30°C temperature, two areas are observed: of decreasing and increasing π/c. Higher values were noted for the 40°C temperature, which could be explained by a change of the polymer chain conformation from helix to coil. Helical structures exhibit high rigidity which hinders the movement of the solvent in the solution. This behaviour changes due to increased temperature and the helix-coil transformation. The noted values of π/c were contained in the range from 10 to 40 $mmH_2O/g\ dL^{-1}$. The virial coefficients, as for all the other cases, are negative. At 40°C, the value of A_2 is 10 times higher than at 30°C, which confirms the influence of temperature on the Xanthan gum chain solvation interactions.

The correlation of π/c and concentration obtained for the konjac gum is descending at both temperatures. The tests were carried out at concentrations below c^* (**Table 5**). The values of π/c obtained at 40°C temperature are approximately two times lower than at 30°C. This indicates that an increase of temperature does not increase the affinity of konjac gum chains to the

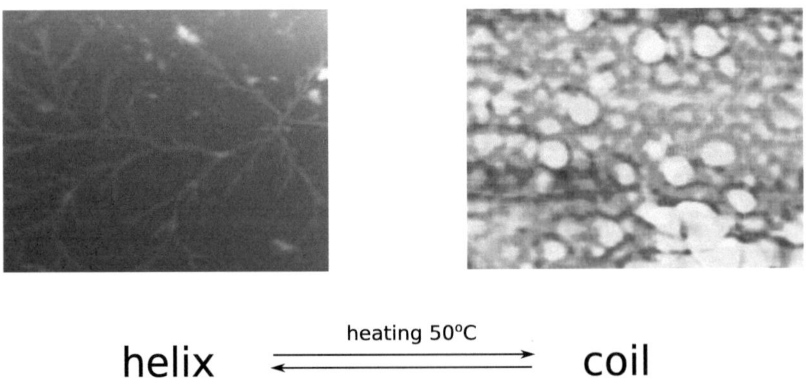

Figure 12. Structure changes of xanthan gum chains in water solutions, AFM images (own study, GONOTEC Qscope, USA, non-contact mode).

http://dx.doi.org/10.5772/intechopen.69864

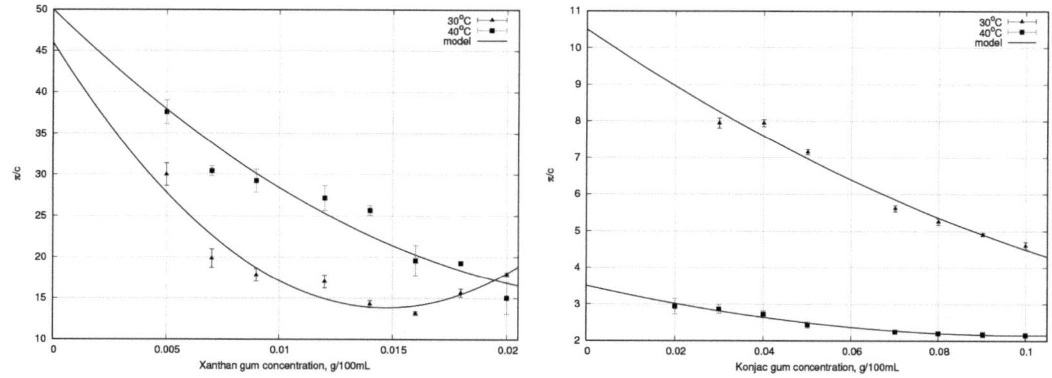

Figure 13. Correlation of the reduced osmotic pressure π and the concentration of xanthan gum (left) and konjac (right) water solutions in the function of concentration (measurements at 30 and 40°C).

solvent. This situation can be due to high viscosity of this polysaccharides solutions (viscosity effect), which hinders the movement of solvent in the solution and by the presence of protein in the konjac gum preparations, which can hinder contact between the solvent and the polysaccharide. Moreover, such behaviour can be a result of this polysaccharides tendency to gelling. Negative virial coefficient confirms low affinity of polymer chains to the solvent.

In the case of carboxymethylcellulose (CMC) (**Table 5**) significantly different behaviours are observed, than in the above-mentioned examples, as the obtained correlation is ascending in the whole range of concentrations (0.005–0.02 g/100 mL) and temperatures (**Figure 14**). Slow increase of the reduced osmotic pressure is a result of high viscosity of the solution, which hinders the migration of the solvent in the solution. High values of π/c (up to 250 mm H_2O/g dL^{-1}) indicate that the CMC chains have the ability to absorb water. This is closely related to the CMC chains conformation in water, where they are completely straight. Such system results in the highest hydrodynamic radius and facilitation of solvation. An increase

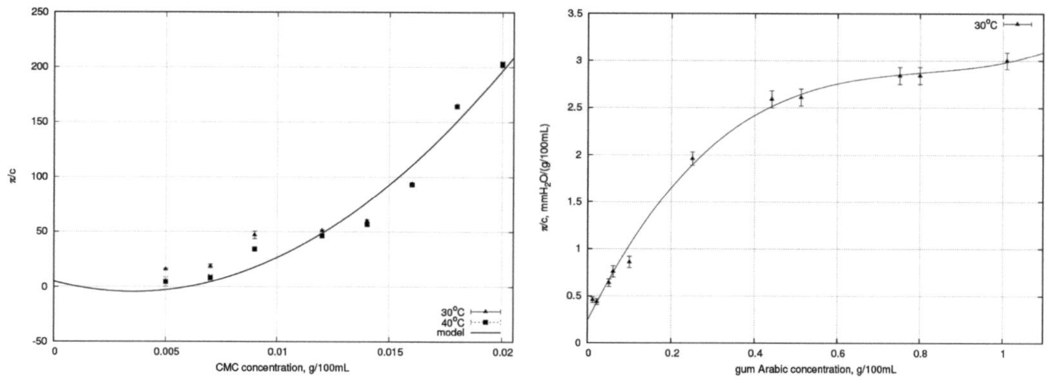

Figure 14. Correlation of the reduced osmotic pressure π and the concentration of carboxylmethylcelullose (left) and gum arabic (right) water solutions in the function of concentration (measurements at 30 and 40°C).

of temperature causes slight changes in the value of π/c reduced osmotic pressure. The second virial coefficient at the tested temperatures is negative and assumes the same value for both 30 and 40°C, which confirms the lack of influence of temperature on the osmotic properties of the studied CMC solutions. It should be noted that the value of the second virial coefficient is the lowest among all the tested hydrocolloids, where its value was negative; this is due to the viscosity effect.

For gum arabic, the relation of π/c to the concentration is ascending (**Figure 14**), similarly to the CMC solutions. However, an increase of gum arabic causes a much faster increase of the π/c value than for the CMC solutions. These observations are closely related to the structure of the gum arabic chain, which is formed of a protein chain with attached polysaccharide chains. Such structure results in a large contact area of polysaccharide chains and the solvent. Moreover, the low viscosity, compared to CMC solutions, does not prevent migration of the solvent in the solution. The calculated value of the second virial coefficient is positive, which enables us to assume, that strong interactions occur between the biopolymer and the solvent and therefore, water is a good solvent for this type of systems [85, 86].

The π/c correlation obtained for carrageenan water solutions is descending in the whole range of concentrations (**Figure 15**). From the concentration of 0.25 g/100 mL, the decrease of the π/c value is less dramatic than for lower concentrations [88]. A negative second virial coefficient would suggest a low affinity of the solvent to the polymer chains, which is due to the low temperature of the osmotic pressure measurement.

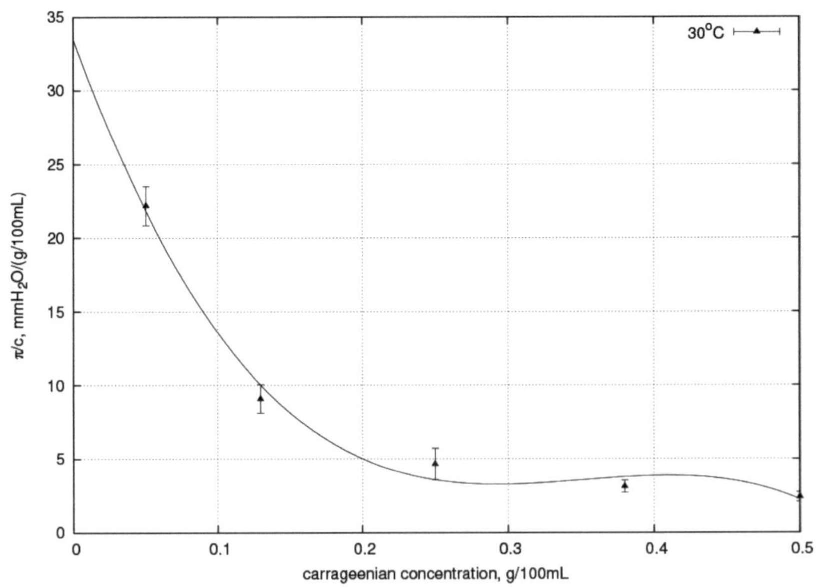

Figure 15. Correlation of the reduced osmotic pressure π and the concentration of carrageenan water solutions in the function of concentration (measurements at 30°C).

Author details

Kruk Joanna, Pancerz Michał and Ptaszek Anna*

*Address all correspondence to: a.ptaszek@ur.krakow.pl

Department of Engineering and Machinery for Food Industry, Faculty of Food Technology, Agriculture University in Krakow, Kraków, Poland

References

[1] Eliasson A-C, editor. Starch in Food: Structure, Function and Applications. Cambrigde: Woodhead Publishing Limited; 2004

[2] Aberle T, Burchard W, Vorwerg W, Radosta S. Conformational contributions of amylose and amylopectin to the structural properties of starches from various sources. Starke. 1994;**46**:329-335. DOI: 10.1002/star.19940460903

[3] Aberle T, Burchard W. Starches in semidilute aqueous solution. Starke. 1997;**49**:215-224. DOI: 10.1002/star.19970490602

[4] Ghosh S, Chattoraj S, Chowdhury R, Bhattacharyya K. Structure and dynamics of lyzozyme in DMSO-water binary mixture: Fluorescence correlation spectroscopy. RSC Advances 2014;**4**:14378-14384. DOI:10.1039/C4RA00719K

[5] Heineck ME, Cardoso MB, Giacomelli FG, da Silveira NP. Evidences of amylose coil-to-helix transition in stored dilute solutions. Polymer. 2008;**49**:4386-4392. DOI: 10.1016/j.polymer.2008.07.062

[6] Miles MJ, Morris VJ, Orford PD, Ring SG. The roles of amylose and amylopectin in the gelation and retrogradation of starch. Carbohydrate Research. 1985;**135**:271-281. DOI: 10.1016/S0008-6215(00)90778-X

[7] Tester RF, Morrison WR, Swelling and gelatinization of cereal starches. I. Effects of amylopectin, amylose and lipids. Cereal Chem. 1990;**67**:551-557

[8] Radosta S, Haberer M, Vorwerg W. Molecular characteristics of amylose and starch in dimethyl sulfooxide. Biomacromolecules. 2001;**2**:970-978. DOI: 10.1021/bm0100662

[9] Ring SG, L'Anson KJ, Morris VJ. Static and dynamic light scattering studies of amylose solutions. Macromolecules. 1985;**18**:182-188. DOI: 10.1021/ma00144a013

[10] Bello-Perez LA, Roger P, Baud B, Colonna P. Macromolecular features of starches determined by aqueous high-performance. Size exclusion chromatography. Journal of Cereal Science. 1998;**27**(3):267. DOI: 10.1094/CCHEM.1998.75.4.395

[11] Roy S, Banerjee S, Biyani N, Jana B, Bagchi B. Theoretical and computational analysis of static and dynamic anomalies in water-DMSO binary mixture at low DMSO concentration. J. Phys. Chem. B. 2011;**115**:685-692. DOI: 10.1021/jp109622h PMID: 21186810

[12] Jordan RC, Brant DA. Unperturbed dimensions of amylase in binary water/dimethyl sulfoxide mixtures. Macromolecules. 1980;**13**:491-499. DOI:10.1021/ma60075a006

[13] Cheetham NWH, Tao L. Amylose conformational transitions in binary DMSO/water mixtures. Carbo-hydr. Polym. 1998;**35**:287-295. DOI: 10.1016/S0144-8617(97)00151-3

[14] Divers T, Balnois E, Feller JF, Spevacek J, Grohens Y. The influence of O-formylation on the scale of starch macromolecules association in DMSO and water. Carbohydr. Polym. 2007;**68**:136-145. DOI: 10.1016/j.carbpol.2006.07.026

[15] De Vasconcelos CL, de Azevedo FG, Pereira MR, Fonseca JLC. Viscosity-temperature-concen-tration relationship for starch-DMSO-water solutions. Carbohydr. Polym. 2000;**41**:181-184. DOI: 10.1016/S0144-8617(99)00078-8

[16] Han JA, Lim ST. Structural changes of corn starches by heating and stirring in DMSO measured by SEC-MALLS-RI system. Carbohydr. Polym. 2004;**55**:265-272. DOI: 10.1016/j.carbpol.2003.09.00

[17] Chakraborty S, Sahoo B, Teraoka I, Gross RA. Solution properties of starch nanoparticles in water and DMSO as studied by dynamic light scattering. Carbohydr. Polym. 2005;**60**:475-481. DOI: 10.1016/j.carbpol.2005.03.011

[18] Yang C, Meng B, Chen M, Liu X, Hua Y, Ni Z. Laser-light-scattering study of structure and dynamics of waxy corn amylopectin in dilute aqueous solution. Carbohydrate Polymers. 2006;**64**:190-196. DOI: 10.1016/j.carbpol.2005.11.017

[19] Ptaszek P, Lukasiewicz M, Ptaszek A, Grzesik M. Rheological scaling properties of starch solutions in dimethylsulfoxide. Chemical and Process Engineering. 2012;**33**:323-333. DOI: 10.2478/v10176-012-0029-7

[20] Ptaszek P, Lukasiewicz M, Ptaszek A, Grzesik M. Rheological scaling properties of starch solu-tions in dimethylsulfoxide. Chemical and Process Engineering. 2012;**33**:323-333. DOI: 10.2478/v10176-012-0029-7

[21] Ptaszek A, Ptaszek P, Dziubiński M, Grzesik N M, Liszka-Skoczylas M. The effect of structural properties on rheological behaviour of starches in binary dimethyl sulfoxide-water solutions. Plos One. 2017;**2**:1-15. DOI:10.1371/journal.pone.0171109

[22] Żmudziński D, Ptaszek A, Grzesik M, Kruk J, Kaczmarczyk K, Liszka-Skoczylas M. Influence of starch acetylation on selected rheological properties of pastes. Starch-Starke. 2014;**66**:303-315. DOI: 10.1002/star.201300094

[23] Dakia PA, Blecker C, Robert C, Wathelet B, Paquot M. Composition and physicochemical properties of locust bean gum extracted from whole seeds by acid or water dehulling pre-treatment. Food Hydrocolloids. 2008;**22**:807-818. DOI: 10.1016/j.foodhyd.2007.03.007

[24] Ensminger AH, Ensminger ME, Konlande JE, Robson JRK. Food and Nutrition Encyclopedia. 2nd ed. Vol. 1. Boca Raton: CRC Press LLC; 1994. pp. 346-348

[25] Daas PJH, Schols HA, de Jongh HHJ. On the galactosyl distribution of commercial galactomannans. Carbohydrate Research. 2000;**329**:609-619. DOI:10.1016/S0008-6215(00)00209-3

[26] McCleary BV, Matheson NK. α-D-galactosidase activity and galactomannan and galactosyl-sucrose oligosaccharide depletion in germinating legume seeds. Phyfochemistry. 1974;**13**:1747-1757. DOI: 10.1016/0031-9422(74)85084-3

[27] da Silva IAL, Gomçalves MP. Studies on a purification method for locust bean gum by precipita-tion with isopropanol. Food Hydrocolloids. 1990;**4**(4):277-287. DOI: 10.1016/S0268-005X(09)80204-X

[28] Lazaridou A, Biliaderis CG, Izydorczyk MS. Structural characteristics and rheological proper-ties of locust bean galactomannans: a comparison of samples from different carobtree populations. J Sci Food Agric. 2000;**81**:68-75. DOI: 10.1002/1097-0010(20010101)81:13.0.CO;2-G

[29] Dea ICM, Morrison A. Chemistry and interactions of seed galactomannans. Adv. Curbohydr. Chem. Biochem. 1975;**31**:241-312. DOI: 10.1016/S0065-2318(08)60298-X

[30] Daas PJH, Schols HA, de Jongh HHJ. On the galactosyl distribution of commercial galac-tomannans. Carbohydrate Research. 2000;**329**:609-619. DOI:10.1016/S0008-6215(00)00209-3

[31] Gainsford SE, Harding SE, Mitchell JR, Bradley TD. A comparison between the hot and cold water-soluble fractions of two locust bean gum samples. Carbohydrate Polymers. 1986;**6**:423-442. DOI: 10.1016/0144-8617(86)90002-0

[32] McClary BV.The fine structures of carob and guar galactomannans. Carbohydrate Research. 1985;**139**:237-260. DOI: 10.1016/0008-6215(85)90024-2

[33] Körk MS. A comparative study on the compositions of crude and refined locust bean gum: In re-lation to rheological properties. Carbohydrate Polymers. 2007;**70**(1):68–76. DOI:10.1016/j.carbpol.2007.03.003

[34] Richardson PH, Willmer J, Foster TJ. Dilute solution properties of guar and locust bean gum in sucrose solutions. Food Hydrocolloids. 1998;**12**:339-348. DOI: 10.1016/S0268-005X(98)00025-3

[35] Körk MS, Hill SE, Mitchel JR. A comparison of the rheological behaviour of crude and refined locust bean gum preparations during thermal processing Carbohydrate Polymers 1999;**38**(3):261-265. DOI:10.1016/S0144-8617(98)00100-3

[36] Simões J, Nunes FM, Rosário Domingues M, Coimbra MA. Demonstration of the presence of acetylation and arabinose branching as structural features of locust bean gum galactomannans. Carbohydrate Polymers. 2011;**86**:1476-1483. DOI: 10.1016/j.carbpol.2011.06.049

[37] Rizzo V, Tomaselli F, Gentile A, La Malfa S, Maccarone E. Rheological properties and sugar composition of locust bean gum from different carob varieties (Ceratonia siliqua L.). J. Agric. Food Chem. 2004;**52**:7925-7930. DOI: 10.1021/jf0494332

[38] Goycoolea FM, Morris ER, Gidley MJ. Viscosity of galactomannans at alkaline and neutral pH: evidence of 'hyperentanglement' in solution. Carbohydrate Polymers. 1995;**27**(1): 69-71. DOI:10.1016/0144-8617(95)00030-B

[39] Crescenzi V, Dentini M, Risica D, Spadoni S, Skjåk-Braek G, Capitani D, Mannina L, Viel S. C(6)-oxidation followed by C(5)-epimerization of guar gum studied by high field NMR. Biomacromolecules. 2004;**5**:537-546. DOI: 10.1021/bm034387k

[40] Grasdalen H, Painter TJ. NMR studies of composition and sequence in legume seed galac-tomannans. Carbohydr Res. 1980;**81**:59-66. DOI: 10.1016/S0008-6215(00)85677-3

[41] Shimahara H, Suzuki H, Sugiyama N, Nisizawa K. Isolation and characterization of oligosaccharides from an enzymic hydrolysate of konjac glucomannan. Agricultural and Biological Chemistry. 1975;**39**(2):293-299. DOI: 10.1080/00021369.1975.10861604

[42] Maeda M, Shimahara H, Sugiyama N. Detailed examination of the branched structure of konjac glucomannan. Agricultural and Biological Chemistry. 1980;**44**(2):245-252. DOI: 10.1080/00021369.1980.10863939

[43] Bourriot S, Garnier C, Doublier JL. Phase separation, rheology and microstructure of micellar casein-guar gum mixtures. Food Hydrocolloids. 1999;**13**:43-49. DOI: 10.1016/S0268-005X(98)00068-X

[44] Crescenzi V, Dentini M, Risica D, Spadoni S, Skjåk-Braek G, Capitani D, Mannina L, Viel S. C(6)-oxidation followed by C(5)-epimerization of guar gum studied by high field NMR. Biomacro-molecules. 2004;**5**:537-546. DOI: 10.1021/bm034387k

[45] Barbucci R, Pasqui D, Favaloro R, Panariello G. A thixotropic hydrogel from chemically cross-linked guar gum: synthesis, characterization and rheological behavior. Carbohydrate Research. 2008;**343**(18):3058-3065. DOI: 10.1016/j.carres.2008.08.029

[46] Kato K, Matsuda K. Isolation of oligosaccharides corresponding to the branching-point of konjac mannan. Agricultural and Biological Chemistry. 1973;**37**(9):2045-2051. DOI: 10.1080/00021369.1973.10860944

[47] Kato K, Matsuda K. 1969. Studies on the chemical structure of konjac mannan Part I. isolation and characterization of oligosaccharides from the partial acid hydrolyzate of the mannan. Agr. Biol. Chern. 1969;**33**(10):1446-1453. DOI: 10.1080/00021369.1969. 10859484

[48] Shimahara H, Suzuki H, Sugiyama N, Nisizawa K. Isolation and characterization of oligosac-charides from an enzymic hydrolysate of konjac glucomannan. Agr. Biol. Chern. 1975;**39**(2):293-299. DOI: 10.1080/00021369.1975.10861604

[49] Maeda M, Shimahara H, Sugiyama N. Detailed examination of the branched structure of konjac glucomannan. Agric. BioI. Chem. 1980;**44**(2):245-252. DOI: 10.1080/00021369.1980. 10863939

[50] Cescutti P, Campa C, Delben F, Rizzo R. Structure of the oligomers obtained by enzymatic hydrolysis of the glucomannan produced by the plant Amorphophallus konjac. Carbohydrate Research. 2002;**337**(24):2505-2511. DOI: 10.1016/S0008-6215(02)00332-4

[51] Katsuraya K, Okuyama K, Hatanaka K, Oshima R, Sato T, Matsuzaki K. Constitution of kon-jac glucomannan: chemical analysis and 13C NMR spectroscopy. Carbohydrate Polymers. 2003;**53**:83-189. DOI: 10.1016/S0144-8617(03)00039-0

[52] Kato K, Matsuda K. Isolation of oligosaccharides corresponding to the branching-point of konjac mannan. Agr. Biol. Chem. 1973;**37**(9):2045-2051. DOI: 10.1080/00021369.1973.10860944

[53] Tatirat O, Charoenrein S, Kerr WL. Physicochemical properties of extrusion-modi?ed konjac glucomannan. Carbohydrate Polymers. Iowa, USA, 2012;**87**:1545-1551. DOI: 10.1016/j.carbpol.2011.09.052

[54] Ratcliffe I, Williams PA, English RJ, Meadows J. Small stain deformation mesurements of kon-jac glucomannan solutions and influence of borate cross-linking. Carbohydrate Polymers. Cambridge, UK, 2013;**95**:272-281. DOI: 10.1016/j.carbpol.2013.02.024

[55] Chudzik B, Zarzyka B, Śnieżko R. Immunodetection of arabinogalactan proteins in different types of plant ovules. Acta Biologica Cracoviensia Series Botanica. 2005;**47**(1):139-146

[56] Mao CF, Klinthong W, Zeng YC, Chen CH. On the interaction between konjac glucomannan and xanthan in mixed gels: An analysis based on the cascade model. Carbohydrate Polymers. 2012;**89**:98-103. DOI: 10.1016/j.carbpol.2012.02.056

[57] Brenner T, Tuvikene R, Fang Y, Matsukawa S, Nishinari K. Rheology of highly elastic iota-carrageenan/kappa-carrageenan/xanthan/konjac glucomannan gels. Food Hydrocolloids. 2015;**44**:136-144. DOI: 10.1016/j.foodhyd.2014.09.016

[58] Dickinson E. Hydrocolloids at interfaces and the influence on the properties of dispersed systems. Food Hydrocolloids. 2003;**17**:25-39. DOI: 10.1016/S0268-005X(01)00120-5

[59] Akiyama Y, Eda S, Kat¯a K. Gum arabic is a kind of arabinogalactan-protein. Agric. Biol. Chem. 1984;**48**(1):235-237. DOI: 10.1080/00021369.1984.10866126

[60] Yadav MP, Igartuburu JM, Yan Y, Nothnagel EA. Chemical investigation of the structural basis of the emulsifying activity of gum arabic. Food Hydrocolloids.2007;**21**:297-308. DOI: 10.1016/j.foodhyd.2006.05.001

[61] Sharma A, Gautam S, Wadhawan S. Xanthomonas. In: Batt CA, Tortorello ML, editors. Encyclopedia of Food Microbiology. Academic Press is an imprint of Elsevier; 2014. pp. 811-817

[62] Smith F. The constitution of arabic acid. Part I. The isolation of 3-d-galactosido-1-arabinose. J Chem Soc. 1939;744-753. DOI: 10.1039/JR9390000744

[63] Williams PA, Phillips GO 2000. Introduction to food hydrocolloids w: Handbook of hydrocol-loids Edited by Phillips GO, Williams PA. Woodhead Publishing Limited

[64] Chudzik B, Zarzyka B, Śnieżko R. Immunodetection of arabinogalactan proteins in differenttypes of plant ovules. Acta Biologica Cracoviensia Series Botanica. 2005;**47**(1): 139-146

[65] Nieto MB, Akins M. Hydrocolloids in bakery fillings. In: Laaman TR, editor. Hydrocolloids in Food Processing. Blackwell Publishing. Ltd. and Institute of Food Technologists; 2011

[66] Showalter A. M. Arabinogalactan-proteins: structure, expression and function. CMLS Cellular and Molecular Life Sciences. 2001;**58**:139-141. DOI: 10.1007/PL00000784

[67] Islam AM., Phillips GO, Sljivo A, Snowden MJ, Williams PA. A review of recent developmenton the regulatory, structural and functional aspects of gum arabic. Food Hydrocolloids. 1997;**11**(4):493-505. DOI:10.1016/S0268-005X(97)80048-3

[68] Dickinson E. Hydrocolloids at interfaces and the influence on the properties of dispersed sys-tems. Food Hydrocolloids. 2003;**17**:25-39. DOI:10.1016/S0268-005X(01)00120-5

[69] Ratcliffe I, Williams PA, Viebke C, Meadows, J. Physicochemical characterization of konjac glucomannan. Biomacromolecules. 2005;**6**(4):1977-1986. DOI: 10.1021/bm0492226

[70] Harding NE, Ielpi L, Cleary JM. 1995. Genetics and biochemistry xanthan gumproduction by Xanthomonas campestris w: Food Biotechnology Microorganisms. editors Hui YH, Khachatourians GC. VCH Publishers. New York. pp. 495-514

[71] Sharma A, Gautam S, Wadhawan S. Xanthomonas w: Encyclopedia of Food Microbiology Editor-In-Chief Batt CA Editor Tortorello ML Academic Press is an imprint of Elsevier. 2014 Elsevier. 811-817

[72] Jansson PE, Keene L, Lindberg B. Structure of the exocellular polysaccharide from Xanthomonas campestris. Carbohydr Res. 1975;**45**:275-282. DOI: 10.1016/S0008-6215(00) 85885-1

[73] Melton LD, Mindt L, Rees DA, Sanderson GR. Covalent structure of the polysaccharide from Xanthomonas campestris: evidence from partial hydrolysis studies. Carbohydrate Research. 1976;**46**(2):245-257. DOI: 10.1016/S0008-6215(00)84296-2

[74] Al-Assaf S, Phillips GO, Williams PA. Studies on acacia exudate gums. Part I: the molecular weight of Acacia senegal gum exudates. Food Hydrocolloids. 2005;**19**:647-660. DOI: 10.1016/j.foodhyd.2004.09.002

[75] Viebke C, Piculell L, Nilssont S. On the mechanism of gelation of helix-forming biopolymers. Macromolecules. 1994;**27**:4160-4166. DOI: 10.1021/ma00093a017

[76] Dua B, Lia J, Zhanga H, Huangb L, Chenb P, Zhoub J. Influence of molecular weight and degree of substitution of carboxymethylcellulose on the stability of acidified milk drinks. Food Hydrocolloids. 2009;**23**(5):1420–1426. DOI: 10.1016/j.foodhyd.2008.10.004

[77] Shakun M, Maier H, Heinze T, Kilz P, Radke W. Molar mass characterization of sodium car-boxymethyl cellulose by SEC-MALLS. Carbohydrate Polymers. 2013;**95**:550–559. DOI: 10.1016/j.carbpol.2013.03.028

[78] Viebke C, Piculell L, Nilssont S. On the Mechanism of Gelation of Helix-Forming Biopolymers. Macromolecules. 1994;**27**:4160-4166. DOI: 10.1021/ma00093a017

[79] Viebke C, Williams PA. Determination of molecular mass distribution of κ-carrageenan and xanthan using asymmetrical flow field-flow fractionation. Food Hydrocolloids. 2000;**14**:265–270. DOI:10.1016/S0268-005X(99)00066-1

[80] Dea ICM, Morrison A. Chemistry and interactions of seed galactomannans. Advances in Carbohydrate Chemistry and Biochemistry 1975;**31**:241-312. DOI: 10.1016/S0065-2318(08)60298-X

[81] Bourriot S, Garnier C, Doublier JL. Phase separation, rheology and microstructure of micellar casein-guar gum mixtures. Food Hydrocolloids. 1999;**13**:43-49. DOI: 10.1016/S0268-005X(98)00068-X

[82] Barbucci R, Pasqui D, Favaloro R, Panariello G. A thixotropic hydrogel from chemically cross-linked guar gum: synthesis, characterization and rheological behavior. Carbohydrate Research. 2008;**343**(18):3058-3065. DOI: 10.1016/j.carres.2008.08.029

[83] Ratcliffe I, Williams PA, English RJ, Meadows J. Small stain deformation mesurements of konjac glucomannan solutions and influence of borate cross-linking. Carbohydrate Polymers. 2013;**95**:272-281. DOI: 10.1016/j.carbpol.2013.02.024

[84] Cuvelier G, Launay B. Concentration regimes in xanthan gum solutions deduced from flow and viscoelastic properties. Carbohydrate Polymers. 1986;**6**:321-333. DOI: 10.1016/0144-8617(86)90023-8

[85] Smith F. The constitution of arabic acid. Part I. The isolation of 3-d-galactosido-1-arabinose. Journal of the Chemical Society. 1939;744-753. DOI: 10.1039/JR9390000744

[86] Viebke C, Williams PA. Determination of molecular mass distribution of κ-carrageenan and xanthan using asymmetrical flow field-flow fractionation. Food Hydrocolloids. 2000;**14**:265-270. DOI: 10.1016/S0268-005X(99)00066-1

[87] Kishida N, Okimasu S, Kamata T. Molecular weight and intrinsic viscosity of konjac glucomannan. Agric. Biol. Chem. 1978;**42**(9):1645-1650. DOI: 10.1080/00021369.1978.10863226

[88] Kishida N, Okimasu S, Kamata T. Molecular weight and intrinsic viscosity of konjac glucomannan. Agricultural and Biological Chemistry. 1978;**42**(9):1645-1650. DOI: 10.1080/00021369.1978.10863226

[89] Xiaoyan L, Qiang W, Xuegang L, Feng L, Xiaoqing L, Pan H. Effect of degree of acetylation on thermoplastic and melt rheological properties of acetylated konjac glucomannan. Carbohydrate Polymers. 2010;**82**:167-172. DOI: 10.1016/j.carbpol.2010.04.053

Chapter 4

Polysaccharides in Solution: Experimental and Computational Studies

Bimali Jayawardena, Dinesh R. Pandithavidana and
WMC Sameera

Additional information is available at the end of the chapter

http://dx.doi.org/10.5772/intechopen.69863

Abstract

Carbohydrates can be found in many natural sources, and they play a central role in various biological processes. These versatile biopolymers are difficult to dissolve in solutions, and therefore converting them into functional forms is a significant challenge, whereby both experimental and computational studies become critical. This chapter discusses commonly used experimental approaches to increase solubility of carbohydrates in solutions and computational studies to rationalize their conformations and solvation effects. Advances of experimental and computational methods for the study of carbohydrates will guide the design approaches to use carbohydrates in industrially and academically relevant applications.

Keywords: solubility, computational studies, solvent effects, conformations

1. Introduction

Carbohydrates are one of the most abundant biomolecules on the earth, and they involve a number of biological processes, such as energy production and storage, structural maintenance, molecular recognition and cell growth. Polysaccharides are polymers of tens, hundreds or even many thousands of monosaccharides linked together through the glycosidic bonds. They are products of a natural carbon-capture process, photosynthesis, followed by further biosynthetic modifications. Some polysaccharides are produced on a very large scale in the nature, and some have industrial relevance, for example, materials and food applications, either in their native or in chemically modified forms.

Structures of the repeating units of some polysaccharides are shown in **Figure 1** [1]. The natural polysaccharides, cellulose and amylose, for instance, consist of a monosaccharide repeating unit with hydroxyl groups as the only functional group. Other related structures, such as curdlan and inulin, are based on furanosides. Amylopectin has a structure similar to amylose, and these two polysaccharides are the components of starch that consist of a branched structure with many amylose-like chains linked together by (α1–6)-branching points. Dextran is an (α1–6)-linked glucan and holds branching at the secondary hydroxyl groups. For example, in the regular comb dextran, every residue in the backbone is substituted by an (α1–3)-linked glucose unit. Xylans, a component of hemicelluloses, are made up of a (β1–4)-linked xylopyranose backbone, but it can also be branched. Xylose is a pentose, where the pyranose units in xylan do not have a

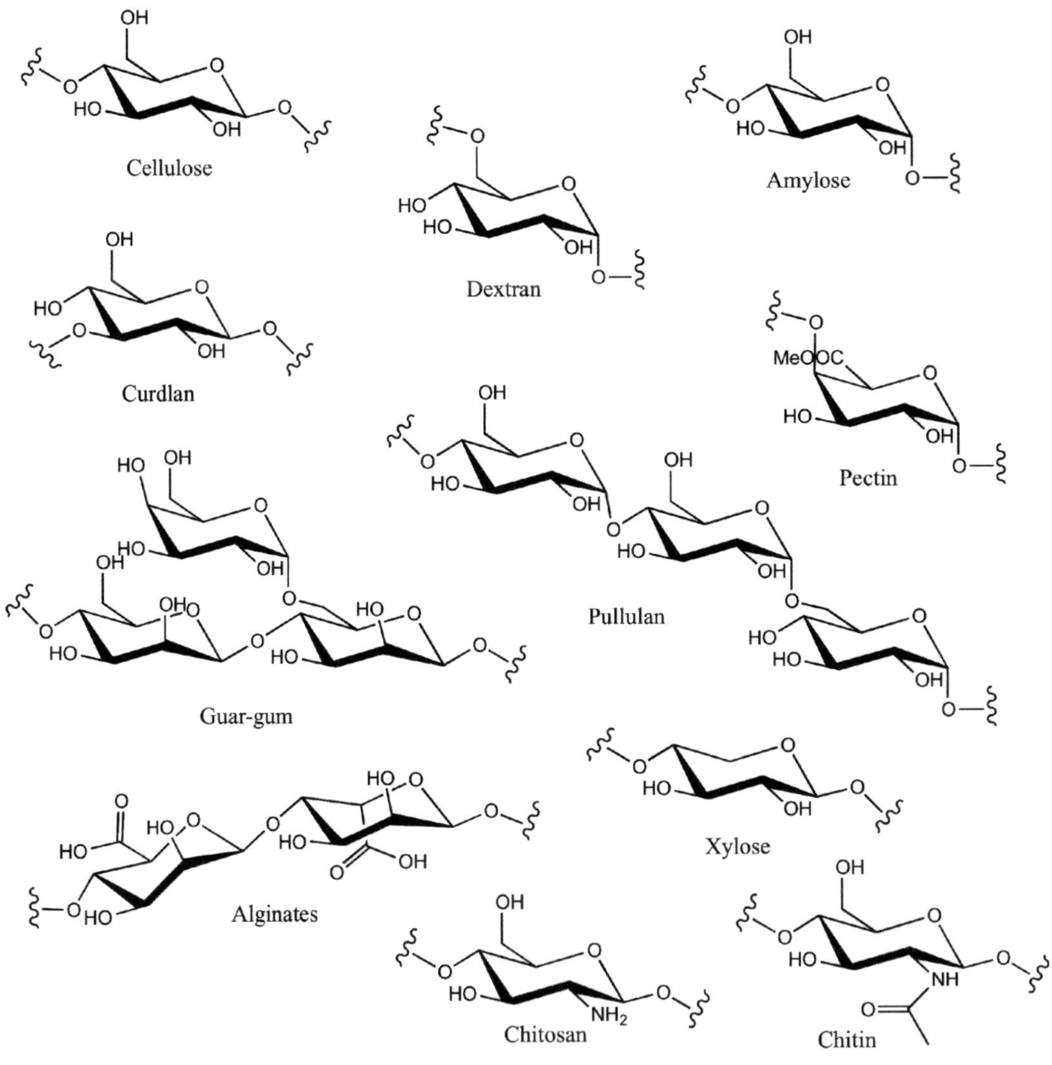

Figure 1. Structures of the repeating units of some of the polysaccharides [1].

primary hydroxyl group. Both guar gum and locust bean gum consist of (β1–4)-linked mannan backbones substituted by Gal(α1–6) units to some extent. In guar gum, approximately every other mannose residue is substituted with galactose, whereas in locust bean gum, long unsubstituted regions alternate with regions of heavy galactose branching [1, 2].

Some polysaccharides have other functional groups as well as the simple hydroxyl groups. Alginates and pectins are based on uronic acids, and their monosaccharide constituents are all oxidized at C-6 to the carboxylic acid level. Alginates consist of domains of (α1–4)-linked L-guluronic acid interspersed with domains of (β1–4)-linked mannuronic acid. Pectins are polysaccharides rich in galacturonic acid, although this acid can be found as its methyl ester. A simple backbone of (α1–4)-linked galacturonic acid methyl ester can also be substituted by other monosaccharide branches. A very common polysaccharide based on aminosugars is chitin/chitosan. The chitin/chitosan relationship can be regarded as a continuum, with polysaccharides containing more of the free base being called chitosan and those mostly N-acetylated being called chitin [3, 4].

The applications of polysaccharides are restricted due to their insolubility in most solvents. Polysaccharides are often insoluble in water or organic solvents. Non-aqueous solvent mixtures that dissolve in cellulose often consist of an organic liquid and an inorganic salt. Examples include DMA (dimethylacetamide)/LiCl, DMF/LiCl, DMI (1,3-dimethyl-2-imidazolinone)/LiCl and DMSO/TBAF (tetrabutylammonium fluoride). The DMSO/Et_3N/SO_2 mixture is a salt-free solvent for cellulose. Sometimes, it is necessary to heat to a high temperature (150°C) before cellulose dissolves in these solvents. Similar solvents or solvent mixtures are often used to dissolve neutral polysaccharides. In general, starch is more soluble than cellulose. Commonly used solvents for chitin include LiCl (5%)/DMA, LiCl/N-methyl-2-pyrrolidone, $CaCl_2$/MeOH and hexafluoroisopropyl alcohol [4–6].

Charged polysaccharides such as chitosan (which may be protonated on nitrogen) or polyuronates such as alginates (which can form carboxylate salts) show very different solubility. Further, chitosan is soluble in aqueous organic or mineral acids below pH 6.5 and also in DMSO [1, 3]. Ionic liquids (ILs) (at room temperature) are relatively new solvents that can be used to dissolve polysaccharides [7], including cellulose, hemicellulose and wood. Cellulose dissolves in ionic liquids, aided by conventional heating, microwave irradiation or sonication, with up to 25% (w/w) being obtained in [bmim]Cl [8]. Other ionic liquids gave 5–10% (w/w) solutions of cellulose. The properties of ionic liquids can be fine-tuned by modifying the structures of two ionic components. Increasing the length of the alkyl chains in the cation component resulted in a less efficient dissolution of cellulose. Amylose shows a very high solubility in ether-derived ionic liquids. Ionic liquids have been called 'green' solvents due to their recyclability and low vapor pressures (low volatility), but a low vapor pressure can limit the recyclability, as its purification becomes difficult. As a result, volatile and distillable ionic liquids have been designed for polysaccharide derivatization [9, 10].

Recently, numerous environmentally friendly methods have been used to dissolve polysaccharides. Subcritical water, microwave digestions and enzyme pretreatments facilitate the dissolution process. However, very little is known regarding the behavior of carbohydrates in solution. Hence, experimental and theoretical characterization of carbohydrate structures is not only critical in terms of establishing the structure-function relationships of carbohydrates but also important in modifying their physical and chemical properties to improve their functional characteristics for numerous technological applications in the food, textile, paper and cosmetics industries. Therefore, both experimental and computational studies play a major role to establish their structural features, to improve their solubility and to design versatile catalysts to convert biomass into other forms such as fuels that are more conveniently used, transported and stored.

2. Carbohydrates in solution

Water is a very strongly hydrogen-bonded liquid, and breaking these bonds requires energy. Solutes that are unable to form hydrogen bonds or decrease the number of hydrogen bonds in water tend to have low solubility. Water-associating polymers and amphiphilic systems with polar units are soluble in water. Dissolution can be improved by modulating thickening, gelling and viscoelasticity through temperature, and in the presence of additives. The characters in the aqueous solutions of such polymers are mainly due to interchain associations between hydrophobic groups leading to physically cross-linked polymer chains.

Most standard techniques used to enhance the rate are by heating and stirring. These techniques speed up and increase the contact between solvents and solute. The kinetic control of dissolution is much more important for macromolecules than for low molecular weight solutes. It is known that multiplicity of inter-chain hydrogen bonds is the cause of aggregation and insolubility of polysaccharides in aqueous solutions. A number of research approaches have been initialized to increase the solubility of polysaccharides in water. The common technique involves the use of different solvents such as alkali or LiCl to break down intermolecular hydrogen bonds of polysaccharides, resulting in dissolution. However, such solvent systems are not applicable for bioactivity assays. Subcritical water (SCW) and super-heated water (SHW) are potential solvents that can be developed to dissolve polysaccharides. SCW has the necessary properties to dissolve water-insoluble polysaccharides as these behave like polar organic solvents and are environmental friendly.

The formation of a homogeneous solution by dissolution of a compound takes place only if the mixed state corresponds to a lower energy than two separate states thermodynamically. Parts of polysaccharide chain become fully solvated by kinetic action mainly by breaking more inter-polysaccharide bonds which are immediately solvated. With time, many sections become solvated and solubilized while fewer amounts of segments are still connected to the other parts of polysaccharide chain. This intermediate stage in the dissolution of the polymer molecule represents a transient gel stage. As the hydration process continues, polysaccharide molecule becomes highly surrounded by partially immobilized water. This layer represents the solvation layer, and molecules move into solution where they may remain monodispersed and preferred low-energy state conformations and shapes where they develop various degrees of gel structure. Polysaccharides dissolve in water by continuous hydration with the transfer of inter-polysaccharide binding to polysaccharide-water binding. The process is facilitated by entropy, as the molecules prefer lower-energy conformations. In all polysaccharides contain sites, where molecules are in a disorganized manner with intermolecular forces and intermolecular hydrogen bonding partially filled because of random spatial arrangement. These amorphous regions therefore have numerous unsatisfied hydrogen-bonding positions that can hydrate [11].

When polymer is in the solvent, solvent molecules rapidly contact the polymer and penetrate into the amorphous region and to the surrounding available polymer sites, competing with and eventually reducing the number of intermolecular bonds, which results in a gel-like consistency. Polymer molecule diffusion is naturally much slower than solvent diffusion and in concentrated polymer solutions, and the diffusion is drastically retarded due to entanglement and aggregation. Entropy is the driving force for dissolution. For spontaneous dissolution, the change in free energy needs to be negative. The reason for difficulty of dissolving polymers is due to higher the molecular weight weaker the entropic-driving force of dissolution.

3. Methods for dissolving carbohydrates

The multiplicity of inter-chain hydrogen bonds is the origin of the aggregation and insolubility of polysaccharides in aqueous solution. Various methodologies have been embraced to decrease this impact [12]. A typical method includes the use of solvent such as alkali or LiCl/Me_2SO to break the intermolecular hydrogen bonds of the polysaccharides, resulting in dissolution. However, such solvent systems are not applicable for bioactivity assays [12]. Apart from that, various chemical modifications have been developed to enhance the solubility of polysaccharides (**Figure 2**). One such chemical modification is sulphation. According to the literature, it is prominent that water-insoluble polysaccharides indicate lower bioactivity. However, their sulphated derivatives show higher solubility that allows them to provide better functional attributes. So here, the sulphation has altered the bioactivity of polysaccharide by changing its chain conformation [13]. Molten salts with low melting points are known as ionic liquids (ILs). In recent years, ILs have been proposed as solvents for polysaccharides. According to Passos and Coimbra [14], 1-n-butyl-3-methylimadiazolium chloride is capable of dissolving cellulose. Due to the high melting temperatures and high viscosities of chloride salts, energy is needed in both the pretreatment and main processes. However, they suggest that carboxylic salt-type ILs are better solvents for a series of polysaccharides [15].

Physical methods (**Figure 2**) have been produced to disperse the supramolecular aggregates by enhancing the energy of the polymer chains. Through these methodologies, a homogeneous aqueous solution of polysaccharides can be prepared. The microwave heating strategy is a feasible method for solubility of the water-insoluble polysaccharides using only compressed water or dilute aqueous solutions. This strategy is also known as the 'microwave superheated water extraction' of polysaccharides and reported to permit milder response conditions, low production costs, arrangement of cleaner items with higher yields and minor wastes when contrasted to other methodologies [14]. When utilizing microwave-assisted extraction, temperature is one of the most important factors contributing to the recovery yield. The presence of higher temperature, higher recovery yield can be obtained. However, these high temperatures can cause degradation of products.

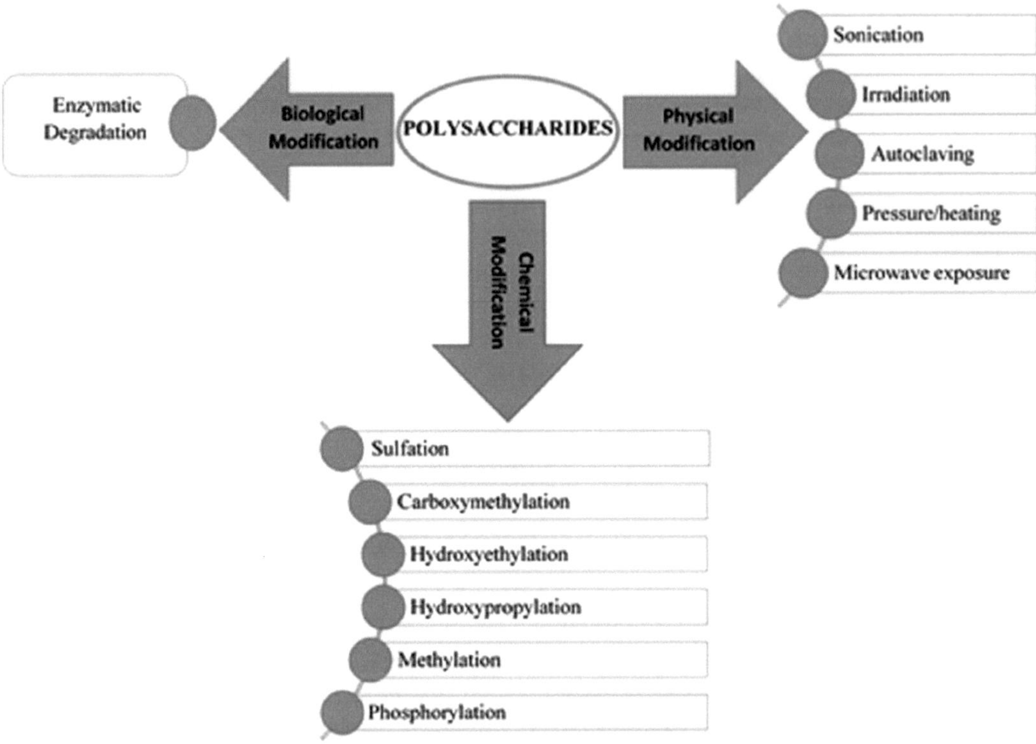

Figure 2. Methods for dissolving carbohydrates.

Ultrasonication is another physical approach used to dissolve polysaccharides. In this process, high molecular weight polysaccharide chains degrade to smaller sizes upon exposure to ultrasound radiation. Despite the fact that the exact mechanism is unclear, it has been attributed to cavitation, the formation and collapse of microscopic vapor bubbles produced by strong sound waves. The resultant stun waves give energy to the polysaccharide, which results in scission if polysaccharide is unable to disperse this energy.

4. Computational studies

In order to understand structure-function relationships of carbohydrates, the first step would be to understand conformations of their monomer units. A fundamental problem in the study of carbohydrates is the extent of conformational complexity, arising from the multitude of bonding possibilities for the primary and secondary hydroxyl groups, the ring-puckering modes and the rotational flexibility of the glycoside linkages. For example, a pyranoseanomer, there are five hydroxyl and one hydroxymethyl rotational dihedral angles, each of which has three staggered conformations. As a result, 729 different stereoisomers are possible. The number of conformations becomes 2916 when both two anomeric and the two main pyranose chair forms are taken into account. When we consider dimers or trimers, the total number of conformation becomes enormous. At the same time, interactions between monomer units give rise to additional complexity. The aqueous phase investigation can be focused on deducing the interactions between polysaccharides and water with extending the chain length of the polymer, varying branching patterns and the substituent position of carbohydrate molecules [16, 17, 18].

The spectroscopic methods can be used for the study of carbohydrate conformations with some success. For example, X-ray crystallography and nuclear magnetic resonance (NMR) spectroscopy provide some insights about the carbohydrate conformationals. The electron microscopy, light or neutron diffraction, and infrared spectroscopy can also provide useful information [19]. Stability of a particular conformation can be determined by the balance between covalent and non-covalent interactions. These interactions determine the structure-function relationships of carbohydrates. This is indeed difficult to characterize in full from experimental studies, and therefore computational studies become critical.

Computational methods can be used to understand carbohydrate conformations. Commonly used semi-empirical methods, MNDO, AM1 and PM3 for instance, fail to describe weak interactions of carbohydrate conformations due to poor dispersion effects. On the other hand, force field methods (e.g., MM3, CHARMM, AMBER, GLYCAM, OPLS and GROMOS) performed well. Therefore, force field methods are commonly used molecular dynamics (MD) simulation of large carbohydrate structures.

Quantum chemical methods have been used to mono- and disaccharide conformations. However, quantum chemical methods are difficult to apply for very large carbohydrate structures due very high computational cost. The Hartree-Fock (HF) or standard density functional theory (DFT) describes hydrogen bonding, but these methods are failed to describe dispersion effects. According to recent studies, non-covalent interactions of biomolecules can be described accurately by using coupled cluster theory using single, double and perturbative triple excitations, CCSD(T), Møller-Plesset (MP) perturbation theory levels and variants, such as spin component-scaled (SCS) approach and the scaled opposite spin (SOS) approach. Also, local pair natural orbital (LPNO) techniques and coupled-electron pair approximation (CEPA) perform well for non-covalent interactions. These ab initio methods are, however, computationally demanding, and therefore some approximation methods, dispersion corrections for density functional and symmetry-adapted perturbation theory (DFT-SAPT) for instance, are viable alternatives. The quantum mechanics/molecular mechanics (QM/MM) methods can be applied for large systems, where non-covalent interactions are described in force field parameterization.

Over the past years, computational methods have been used for the study of monosaccharides and provided very useful information about the accuracy of modern computational approaches in describing the energies of small saccharides [20–26]. A recent study rationalized a comprehensive benchmark study of a test set of 58 structures [27]. In this work, coupled-cluster calculations extrapolated to the complete basis set limit (CCSD(T)/CBS) was used as reference energy and compared with the Møller-Plesset perturbation theory and its variants, the localized paired natural orbital-coupled electron pair approach (LPNO-CEPA) and 31 DFT methods (**Figure 3**). This study suggested that the LPNO-CEPA provides CCSD(T) quality results, and MP2 and SCS-MP2 also give good results. Among the density functional methods, the mPW2PLYP-D and M06-2X are the best choices.

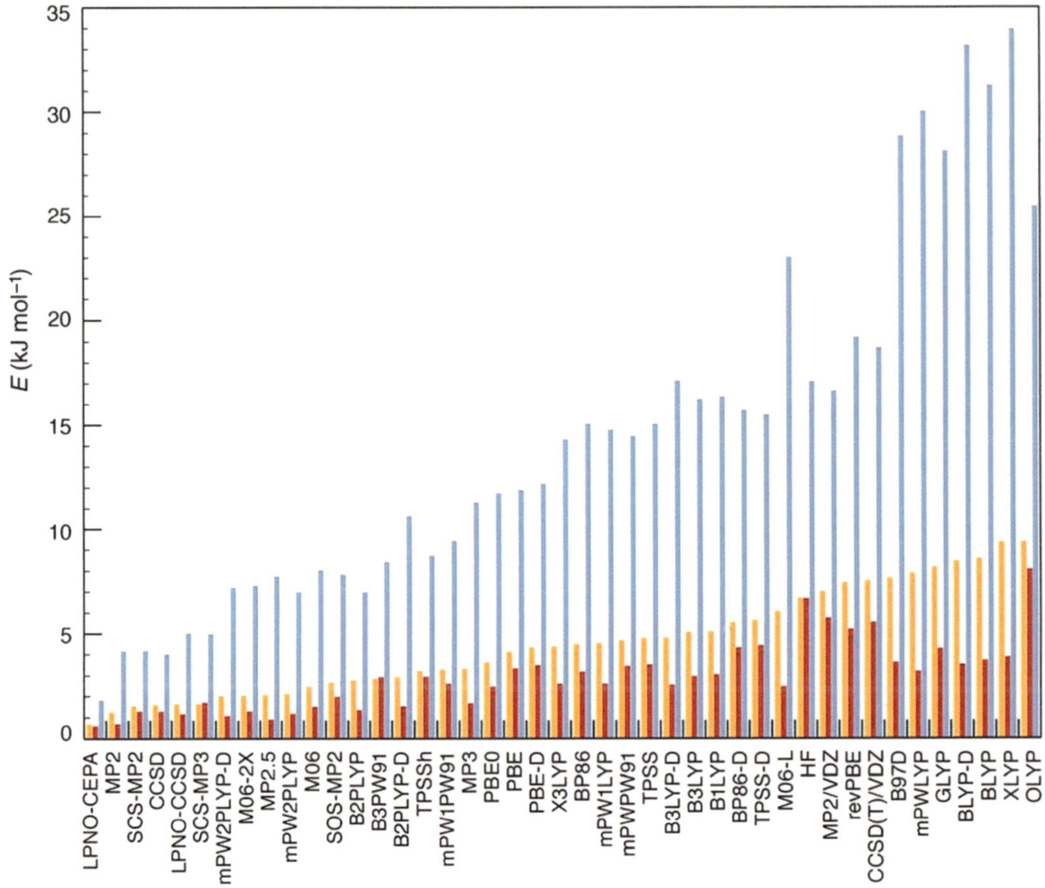

Figure 3. Calculated total average error, average errors excluding open-chain isomerism and maximum absolute errors. Reprinted with adaptation by permission from Ref. [27]. Copyright 2008 American Chemical Society.

Application to larger systems, such as plant or bacterial polysaccharides, will necessitate advances in not only computer performance but also theoretical approaches. Conformational sampling over a nanosecond time scale, which is the current limit of solvated molecular dynamics simulations, is still far too short to adequately sample the conformational space available to a polysaccharide in solution. MD simulations may achieve considerably more sampling, and the absence of explicit water is in some way compensated for within the force field, which may be applicable to polysaccharides [28]. Crystalline environments may be adequately sampled on much shorter time scales, and MD simulations of solid-state polysaccharides are useful in their structural and energetic interpretations [29]. Alternative techniques, such as Langevin dynamics, have been applied elegantly in macroscopic simulations of DNA super coiling, involving as many as 1000 base pairs, and may be applicable to polysaccharide simulations. However, preliminary Langevin dynamics on small molecules related to sugars indicate that it may be necessary to include at least a first shell of explicit waters [30]. Longer time-scale motions may also be approached by Brownian dynamics. In these simulations, the level of detail has to be sacrificed; however, the benefits of obtaining a macroscopic model on the micro- or even millisecond time scale would be significant [30, 31].

Several different approaches have been suggested in literature to model solvent effects on these biomolecules [32, 33]. One possible way to overcome part of this problem is to include a limited number of solvent molecules in the calculation. In this case, the system of the solute and a small number of solvent molecules is treated as a supermolecule. The supermolecule approach may be useful in determining the specific solvation sites.However, for most processes in saccharide solution, the overall solvation energy rather than the optimal solvation sites is of interest. An alternative approach is to treat the solvent as a dielectric continuum (i.e. implicit solvation). This approach was based on the solvophobic model and was developed for saccharides [18].The continuum models are useful in estimating conformational energies of polar molecules. However, they are of limited validity for the quantitative treatment of ionic reactions and for studies of the balance of interactions between the solute-solute and solute-solvent.

A combination of supermolecule and continuum approaches has been reported recently. Probably, ultimate approach for studying solvent effects will be the brute force method of computer simulation of the given solution. This can be based on either Monte Carlo or molecular dynamics calculations including explicit solute and many solvent molecules. However, they are not suitable potential functions for calculating the potential energy of saccharides in different solvents. The method of free-energy calculation of possible conformers in dilute solutions involves creating a cavity of sufficient size to accommodate the solute molecule. The cavity formation requires a Gibbs free energy (ΔG_{cav}). Firstly, the solute molecule is introduced to the designed cavity, which then interacts with the surrounding solvent molecules. The calculation of the cavity term is based on an expression taken from the scaled particle theory, which has been successfully used for the study of thermodynamic properties of aqueous and non-aqueous solutions [34]. The electrostatic term is calculated according to theory of Onsager reciprocal relations of the reaction field that expresses the equality of certain ratios between flows and forces in thermodynamic systems out of equilibrium, but where a notion of local equilibrium exists as applied by Abraham and Bretschneider. The dispersion interactions take into account both attractive and repulsive non-bonded interaction, using a combination of the London dispersion equation and Born-type repulsions [35].

As a simple parameter for visualizing linkage dynamics and water interaction, intramolecular hydrogen bonds can be investigated computationally at regular intervals throughout the cellulose simulation. Their definition is based on bond angle and bond distance criteria. According to the literature, the number of observed hydrogen-bond pairs has been calculated as a function of time [36]. After a period as short as 500 ps, all of the persistent intramolecular hydrogen bonds can be observed, implying that these conformations are highly sampled during the simulation. Furthermore, few new interactions are seen after a period of 2 ns had elapsed, indicating that a large percentage of conformational space has been sampled during this period and confirming that simulations of length 5 ns are adequate to ensure statistical sampling of the major conformers. This is also evident from the observation that each of the internal linkages in the decasaccharide populated space in an identical way. If this is not true, then inadequate conformational sampling would be suspected. Similar checks can made probably for all of the decasaccharide simulations, with similar results.

5. Conclusions

In this chapter, we discussed recent advances in experimental and theoretical studies of polysaccharides in solution. Solubility of polysaccharides in solution is very low, and therefore the development of suitable solvents for polysaccharides becomes critical in terms of converting them into other functional forms. To understand structure-function relationships of polysaccharides, computational studies are very important, and theory rationalizes the low-energy conformation of monomer, dimer or trimer units of polysaccharides. However, low-energy conformations of longer chains and interactions between solvent molecules are significant challenges for computational methods. MD simulations are well suited for such condensed phase modeling. Interactions between polysaccharides and other biological macromolecules are also amenable to computational approaches. Given the sophistication of these methods, it may be tempting to think of the simulation as being truly representative of the properties of the carbohydrate. But many aspects of the real system, such as proton exchange, pH effects, anomerization, induced polarization and other quantum effects, are absent. Despite these limitations, modern computational approaches can provide insight into physical properties, which may not be accessible experimentally.

Author details

Bimali Jayawardena[1], Dinesh R. Pandithavidana[1] and WMC Sameera[2*]

*Address all correspondence to: wmcsameera@sci.hokudai.ac.jp

1 Department of Chemistry, University of Kelaniya, Kelaniya, Sri Lanka

2 Department of Chemistry, Faculty of Science, Hokkaido University, Kita-ku, Sapporo, Hokkaido, Japan

References

[1] Cumpstey I. Chemical Modification of Polysaccharides. ISRN Organic Chemistry. 2013; **2013**:417672-417698. DOI: 10.1155/2013/417672

[2] Yalpani M. Tetrahedron. 1985;**41**:2957-3020. DOI: 10.1016/S0040-4020(01)96652-9

[3] Corma SI, Velty A. Chemical Reviews. 2007;**7**:2411-2502. DOI: 10.1021/cr050989d

[4] Vyver V, Geboers J, Jacobs PA, Sels BF. ChemCatChem. 2011;**3**:82-94. DOI: 10.1002/cctc.201000302

[5] Cunha AG, Gandini A. Cellulose. 2010;**17**:1045-1065. DOI: 10.1007/s10570-010-9435-5

[6] Vigo TL, Sachinvala N. Polymers for Advanced Technologies. 1999;**10**:311-320. DOI: 10.1002/(SICI)1099-1581(199906)10:6<311::AID-PAT880>3.0.CO;2-G

[7] Heinze T, Liebert T. Progress in Polymer Science. 2001;**26**:1689-1762. DOI: 10.1016/S0079-6700(01)00022-3

[8] Gericke M, Fardim P, Heinze T. Molecules. 2012;**17**:7458-7502. DOI: 10.3390/molecules-17067458

[9] Murugesana S, Linhardt RJ. Current Organic Synthesis. 2005;**2**:437-451. DOI: 10.2174/157017905774322640

[10] King AWT, Asikkala J, Mutikainen I, Järvi P, Kilpeläinen I. Angewandte Chemie International Edition. 2011;**50**:6301-6305. DOI: 10.1002/anie.201100274

[11] Lindman B, Karlston G, Stigsson L. Journal of Molecular liquids. 2010;**156**:76-79. DOI: 10.1016/j.ijbiomac.2006.03.030

[12] Tao Y, Xu W. Carbohydrate Research. 2008;**343**:3071-3078. DOI: 10.1016/j.carres.2008.09.009

[13] Nie X, Shi B, Ding Y, Tao W. International Journal of Biological Macromolecules. 2006;**39**:228-233. DOI: 10.1016/j.ijbiomac.2006.03.030

[14] Passos CP, Coimbra MA. Carbohydrate Polymers. 2013;**94**:626-627. DOI: 10.1016/j.carbpol.2013.01.088

[15] Abe M, Fukaya Y, Ohno H. Green Chemistry. 2010;**12**:1274-1275. DOI: 10.1039/C003976D

[16] Tayal A, Khan SA, Macromolecules. 2010;**33**:9488-9489. DOI: 10.1021/ma000736g

[17] Kirschner KN, Woods RJ. Proceedings of the National Academy of Science. 2001;**98**:10541-10545. DOI: 10.1073/pnas.191362798

[18] Cramer CJ, Truhlar DG. Journal of American Chemical Society. 1993;**115**:5745-5753. DOI: 10.1021/ja00066a046

[19] Davis AP, Wareham RS. Angewandte Chemie International Edition. 1999;**3**:2978-2996. DOI: 10.1002/(SICI)1521-3773(19991018)38:20<2978::AID-ANIE2978>3.0.CO;2-P

[20] Wladkowski BD, Chenoweth SA, Jones KE, Brown JW. Journal of Physical Chemistry A. 1998;**102**:5086-5092. DOI: 10.1021/jp980524+

[21] Momany FA, Willett JL. Journal of Computational Chemistry. 2000;**21**:1204-1219. DOI: 10.1002/1096-987X(200010)21:13<1204::AID-JCC9>3.0.CO;2-F

[22] Strati GL, Willett JL, Momany FA. Carbohydrate Research. 2002;**337**:1833-1849. DOI: 10.1529/biophysj.106.081539

[23] Momany FA, Schnupf U, Willett JL, Bosma WB. Structural Chemistry. 2007;**18**:611632. DOI: 10.1007/s11224-007-9191-9

[24] French AD, Johnson GP, Cramer CJ, Csonka GI. Carbohydrate Research. 2012;**350**:68-76. DOI: 10.1016/j.carres.2011.12.023

[25] Csonka GI, French AD, Johnson GP, Stortz CA. Journal of Chemical Theory and Computation. 2009;**5**:679-692. DOI: 10.1021/ct8004479

[26] Csonka GI, Kaminsky JJ. Journal of Chemical Theory and Computation. 2011;**7**:988-997. DOI: 10.1021/ct200007x

[27] Sameera WMC, Pantazis DA. Journal of Chemical Theory and Computation. 2012;**8**:2630-2645. DOI: 10.1021/ct3002305

[28] Stokke BT, Talashek TA, Brant DA. Macromolecules. 1994;**27**:1124-1135. DOI: 10.1021/ma00083a009

[29] Heiner AP, Sugiyama J, Teleman O. Carbohydrate Research. 1995;**273**:207-223. DOI: http://dx.doi.org/10.1016/0008-6215(95)00103-Z

[30] Widmalm G, Pastor RW. Journal of Chemical Society. 1992;**88**:1747-1754. DOI: 10.1039/FT9928801747

[31] Andrew SM, Thomasson KA, Northrup SA. Journal of American Chemical Society. 1993;**115**:5516-5521. DOI: 10.1021/ja00066a020

[32] Northrup SH, Thomasson KA, Miller CM. Biochemistry. 1993;**32**:6613-6623. DOI: 10.1016/S0006-3495(97)78838-6

[33] Jeffrey GA, Nanni R. Carbohydrate Research. 1980;**137**:21-30. DOI: 10.1016/0008-6215(85)85146-6

[34] Tvaroška I, Kožár T. Journal of Quantum Chemistry. 1983;**23**:765-778

[35] Varki A. Current Opinion in Cell Biology. 1992;**4**:257-266. DOI: 10.1016/0955-0674(92)90041-A

[36] Almond A, Sheehan JK. Glycobiology. 2003;**13**:255-264. DOI: 10.1093/glycob/cwg031

Section 3

Common Polysaccharides

Chapter 5

Starch-Galactomannans Mixtures: Rheological and Viscosity Behavior in Aqueous Systems for Food Modeling

Erich von Borries-Medrano,
Mónica R. Jaime-Fonseca and
Miguel Á. Aguilar-Méndez

Additional information is available at the end of the chapter

http://dx.doi.org/10.5772/intechopen.68915

Abstract

Starch properties during processing are major determinants of starch employment for food development. The gelatinization, pasting, and retrogradation of starch can be modified by the addition of galactomannans, which can improve rheological, textural, and nutritional properties. Rheology is an important key to obtain information about thermal transitions of starch and controlling characteristics required in food applications and to enhance understanding of the effect of starch-galactomannans systems and starch properties. This chapter provides information on starch transitions under heating and after cooling, including a definition of the process, molecular mechanisms, and rheological methods and its modification using starch-galactomannans mixtures and explains interactions throughout several investigations. The chapter also discusses how the rheological properties can affect the rate of the enzyme digestibility of starch on *in vitro* measurements and presents the starch-galactomannan systems as an alternative that can be used in structured foods as functional ingredients.

Keywords: starch, galactomannans, rheology, structured food

1. Introduction

Starch is one of the most extensively used and studied biopolymers because of its low cost, accessibility, and ability to provide a wide range of functional properties to food systems. Time-effective structural changes (gel formation processes, retrogradation, and syneresis)

hamper the control of rheological behavior. The retrogradation of starch refers to the process in which disaggregated amylose and amylopectin chains in a gelatinized starch paste reassociate to form more ordered structures [1]. The aggregation and recrystallization of starch molecules are important processes to manage, because these factors contribute to the control of textural properties of starch-based products.

Since native starch pastes and gels often exhibit low resistance to processing conditions, the native starches are replaced by chemically modified starches in food products or mixed with other hydrocolloids. It is well known in the food industry that nonstarch hydrocolloids have many functions, such as improving texture and moisture retention, extending the shelf life of the product, and controlling rheological properties [2, 3]. The control of the rheological properties of starch is significant in order to regulate product processes and to optimize its applicability in food products. Researchers have studied the effect of hydrocolloids on the rheological properties of starches and found that viscosity, the retrogradation of starch dispersions, and syneresis of starch gels are influenced by the addition of hydrocolloids [4–6].

Galactomannans are widely used in the industry due to their functional properties such as thickening, binding, and stabilizing agents. Galactomannans are considered functional ingredients that can change the rate of degradation of carbohydrates such as starch during digestion, thus regulating insulin levels, which is key in the prevention of obesity and diabetes [7].

This chapter was aimed at reviewing detailed recent research on the rheological and functional properties of starches and their mixtures with some galactomannans to develop structured foods. Additionally, the understanding of the rheological characteristics of these products is important to acquire knowledge of these molecular interactions.

Rheological measurements can be important to control the physicochemical properties with the addition of specific carbohydrates such as starch and galactomannans. This research can result in the innovative development of products. Rotational and oscillatory experiments and viscosity behavior are presented as examples.

2. Rheological and viscosity measurement methods

Several techniques are usually employed to characterize the thermal transition of starch and to distinguish between the solid and liquid-like behavior of food. These can be useful in recognizing interactions and structural failures that starch alone can present, supporting the addition of hydrocolloids such as galactomannans, and finding the quantity and type required to enhance the properties of starch.

Pasting can be measured using a Rapid Visco Analyzer (RVA) in order to obtain the pasting profile and the parameters related to this phenomenon. The heating and cooling cycles are programmed to an inherent thermal program in the equipment.

In order to understand the pasting properties of starch after heating, we need to remember that steady shear measurements provide information on the effect of shear in the flow behavior. It can be measured with controlled shear stress or controlled shear rate using a rotational rheometer. Thus, the rotational test is a helpful tool to characterize flow behavior. The flow behavior using the shear stress as a function of shear rate fits into different models, such as power law and Herschel-Bulkey and Bingham models [8]. The data obtained is usually fitted to the well-known power law as follows:

$$\tau = K(\gamma)^{n} \quad (1)$$

where τ is the shear stress (Pa), K is the consistency coefficient (Pa s^n), γ is the shear rate (s^{-1}), and n is the flow behavior index (dimensionless).

The measurement is programmed to a ramp shear rate in two cycles (up curve and down curve) in order to characterize the flow of the paste and determine the hysteresis loop. This area is interpreted as a dimension of energy delivered to the system to break down its thixotropic structure [9]. All liquids with microstructure, like starch, can show thixotropic behavior. Thixotropy reflects the finite time necessary to move from any state of microstructure to another. The hysteresis loop can, to a certain extent, serve as a parameter for quality control. The hysteresis loop can be measured and related to shear rate and time simultaneously [10].

Dynamic rheological measurements reviewed in this chapter are amplitude sweep, frequency sweep, and temperature sweep, and these could be combined to obtain information about structures and interactions. Parameters like G' (storage modulus) and G" (loss modulus) are important in these measurements. Loss tangent (tan δ) is another dynamic rheological parameter that can be used to describe the viscoelastic behavior. This parameter is directly associated with the energy lost per cycle divided by the energy stored per cycle (G'/G''). Values of tan δ <1 and tan δ >1 indicate whether the behavior is elastic or viscous, respectively.

Amplitude sweep tests are done to obtain the strain and to ensure that all dynamic measurements were performed in the linear viscoelastic region (LVE). Amplitude sweep tests can be used to define material structural strength and to distinguish weak and strong gels. A strong gel remains at the LVE region longer compared to a weak gel [11].

The frequency sweep test is done once the LVE region is defined and a strain is fixed, subjecting the samples to oscillatory measurements at a certain frequency range. The information obtained is required to classify the dispersion within this classification: (i) dilute solution, (ii) entanglement network system (concentrated system), (iii) weak gel, and (iv) strong gel [11].

The frequency dependence is generally described by the following power law relationship [12]:

$$\tau = K'\omega^{n'} \quad (2)$$

$$\tau = K''\omega^{n''} \quad (3)$$

where K' and K'' are constants and n' and n'' may be referred to as frequency exponents, and ω is the angular frequency.

The temperature sweep measurements are performed to determine the viscoelastic properties of samples over a certain temperature range (50–95°C), with a strain and frequency defined and a heat rate and the parameters calculated that need to be similar to the RVA parameters [13].

3. Starch

3.1. Starch structure

Starch granules are found in seeds, roots, and tubers and are from different origins such as maize, wheat, potatoes, and rice. Native starch granules present a multilevel structure with at least five levels, ranging in scale from nanometer to micrometer. The size of intact starch granules could be between 1 and 100 μm, while semicrystalline and amorphous growth rings decrease in thickness from 450 to 550 nm near to the core and 80 to 160 nm near to the periphery that are originated most likely during various phases of biosynthesis of starch granules in which the deposition of crystalline layers alternate with amorphous layers. The next level is the blockets that vary from 20 to 500 nm. At a smaller size scale are the left-handed superhelices with their structural elements, crystalline, and amorphous lamellae with a periodicity of 9 nm. The smallest starch structure organization unit is the glucosyl unit (0.3–0.5 nm). Native starch granules present a concentric three-dimensional structure with a total crystallinity between 15 and 45%, depending on the different sources [1].

Starch is mostly composed of two kinds of polysaccharides: amylose and amylopectin. Amylose is a linear α-(1–4) linked glucan, with a molecular weight of 105–106 and a degree of polymerization (DP) as high as 600. Amylose chains can form spiral-shaped single or double helices with a rotation of α-(1–4) link and six glucoses per rotation, and the hydroxyl groups are located toward the exterior of the helices. Otherwise, amylopectin is an α-(1–4) linked glucan with 4.2–5.9% α-(1–6) branch linkages, with constituting branching points every 22–70 glucose units and a high molecular weight of 107–109. This highly multiple structure has an important effect on physical and biological properties. Small amounts of lipids, proteins, and phosphorous, depending of its source, are also found in starch granules [14, 15].

Amylopectin forms double helices, which contribute to the crystalline structure of the granules. Therefore, depending on the branch-chain length (BCL) of amylopectin, native starches granules can present different X-ray diffraction patterns: A-, B-, or C-type. The A-type polymorphic starch consisting of more short branch chains (A and B1 chains) has the double helices packed in a monoclinic unit cell, while B-type starch consists of more long branch-chains (B2, B3, and B4 chains) packed in a hexagonal unit cell. The C-type is a mixture of A- and B-type polymorphs. A-type granules are disk-like or lenticular in shape. On the other hand, B-type starch granules are roughly spherical or polygonal in shape [8].

Potato starch granules (oval and lenticular shapes) are the largest in comparison with wheat (spherical and lenticular shapes), corn (angular shapes), and rice (pentagonal and angular shapes) starches.

3.2. Thermal transitions of starch

Starches are usually processed before eating; therefore, thermal transitions of starch represent a key point in the control of starch functions. The mechanisms involved in these thermal transitions are the same, no matter what the final use is. The associated properties developed on treated starches are related to phase transition, glass transition, physical aging, and to the plasticizing effect of water. Rheological measurements can be used to identify the optimal use of the concentration of carbohydrates, temperature, and heating rate for developing products.

3.2.1. Gelatinization

Gelatinization is presented when native starch granules are heated in the presence of water and is related to an irreversible disruption of their molecular order, semicrystalline structure, and three-dimensional architecture and describes the phase transition from the highly ordered structure of native starch granules to disordered structure in water. During gelatinization several events take place, diffusion of water inside the starch granule with a limited swelling, loss of birefringence, loss of crystallinity of the granule, endothermic phase transitions, granular swelling after loss of birefringence, and molecular solubilization [16].

Small amplitude oscillatory shear (SAOS) measurement is a particularly useful method to study the gelation/gelatinization phenomenon in monitoring the kinetics of network development, as long as the measurements are within the viscoelastic limit, and the strain is restricted to less than 5% [17].

Gelatinization exhibits a two-stage behavior: the first is a limited swelling and a low level of solubilization, which increases while passing through the gelatinization temperature (60–75°C), and the second is soluble polymer molecule leach from the swollen granules that allows these rheological properties as the storage modulus (G') and loss modulus (G'') of starch to increase during change of temperature from 65 to 95°C.

At the beginning of heating, G'' had a higher value than G', reflecting the liquid-like behavior. The crossover temperature in rice starch was in the range of 64.2–77.15°C and represented the gel point [18]. The increase of G' could be attributed to the degree of granular swelling. At the end of the temperature sweep G' was higher than the G'', representing the gel-like behavior that a frequency sweep can support, expressing the formation of a three-dimensional gel network by the intergranular contact of swollen granules [19]. A further increase in temperature led to a G' decrement, indicating that the gel structure was destroyed during prolonged heating as presented in a temperature sweep performed to rice starch [20].

Many factors affect the gelatinization of starch. A positive correlation between the gelatinization temperature of starch and the branch-chain length of amylopectin is that long branch-chains of amylopectin form thermally stable crystallites [21]; therefore, granule botanical source can be important for the development of rheological properties.

The large and cuboidal or irregular-shaped granules in potato starch exhibited higher storage and loss modulus and lower tan δ than small oval granules [22], and the presence of a high phosphate monoester content and the absence of lipids and phospholipids in the potato starch could be also responsible for high G' and G''. The corn starch phospholipids and the more rigid granules could be important in the lower value of G' in corn starch compared with potato starch G', and also, the amylose lipid complex formation during gelatinization of corn starch lowers the G' and G'' [23]. Amylose content enhanced the development of G' and G'' in rice starch [24]. Also, the breakdown, that is, the difference between peak G' (maximum G') at TG′ (onset temperature of G' increasing) and minimum G' was influenced by the granule rigidity and lipid content, and potato starch has a higher breakdown than corn, rice, and wheat starches, which are rich in lipids [23].

3.2.2. Pasting

The second step occurs above the gelatinization temperature. The granules lose their crystalline structure, absorb water, swell, and disrupt. Water controls the formation of starch pastes or gels by forming hydrogen bonds with available hydroxyl groups of starches.

The hot paste created have swollen granules, granule fragments, and a soluble portion (30–60%) [16]. These pasting and thickening properties of pastes are related to involve in applications. Starch paste and gel have rheological properties that depend on different factors, such as starch concentration, pasting conditions (temperature, shear rate, and heating rate), and storage. The viscosity profile of tapioca starch (6%) can be seen in **Figure 1**.

The pasting properties of starch are measured using an amylograph, such as Brabender amylograph and RVA, or using a dynamic rheometer in a flow temperature ramp mode. The dynamic rheology studied starch gelatinization and pasting from a different view from RVA and also requires small deformation on samples. These parameters of RVA measurements are the following:

- Pasting temperature: The point at which the temperature rises above the gelatinization temperature; this leads to starch granules swelling and results in increased viscosity;
- Peak viscosity: The maximum viscosity of material developed after the heating and is indicative of the water-holding capacity of starch;
- Trough viscosity: The lowest viscosity after the peak viscosity;
- Breakdown: The difference between peak viscosity and trough viscosity; represents the disintegration of starch granules under heating due to a rupture of granules and release of soluble amylose;
- Final viscosity: The viscosity value at the test ending;
- Setback: The difference between final viscosity and trough viscosity; is related to re-association between starch molecules, especially amylose; results in a viscosity increment and gel formation, and is related to retrogradation.

Meanwhile, the parameters obtained in a sweep temperature test using dynamic rheology are almost the same [13]:

- Ts: temperature at which complex viscosity (η^*) start increasing
- Tmax: temperature corresponding to maximum η^*
- ηmax: maximum η^* on test
- η^*f: the end value of η^* at 50°C
- ηmin: minimum η^* on test
- Setback: the difference between η^*f and ηmin.

At this stage of heating, the G' and G'' values of starch have reached their maximum values. Thereafter, a plateau (80–85°C) exists, and finally, G' goes under heating and shear time, and the solubilization and swelling have an impact on the drop after the maximum peak. Amylopectin and its branch-chain length are the main components of starch responsible for the swelling power and viscosity development of starch, while amylose and lipids restrict the swelling of starch. Other components of starch are lipids and phosphate-monoester derivatives. Amylose-lipid complex formed in starch under heating develops entanglements with amylopectin molecules and restricts swelling and leads to higher pasting temperature and a lower peak viscosity.

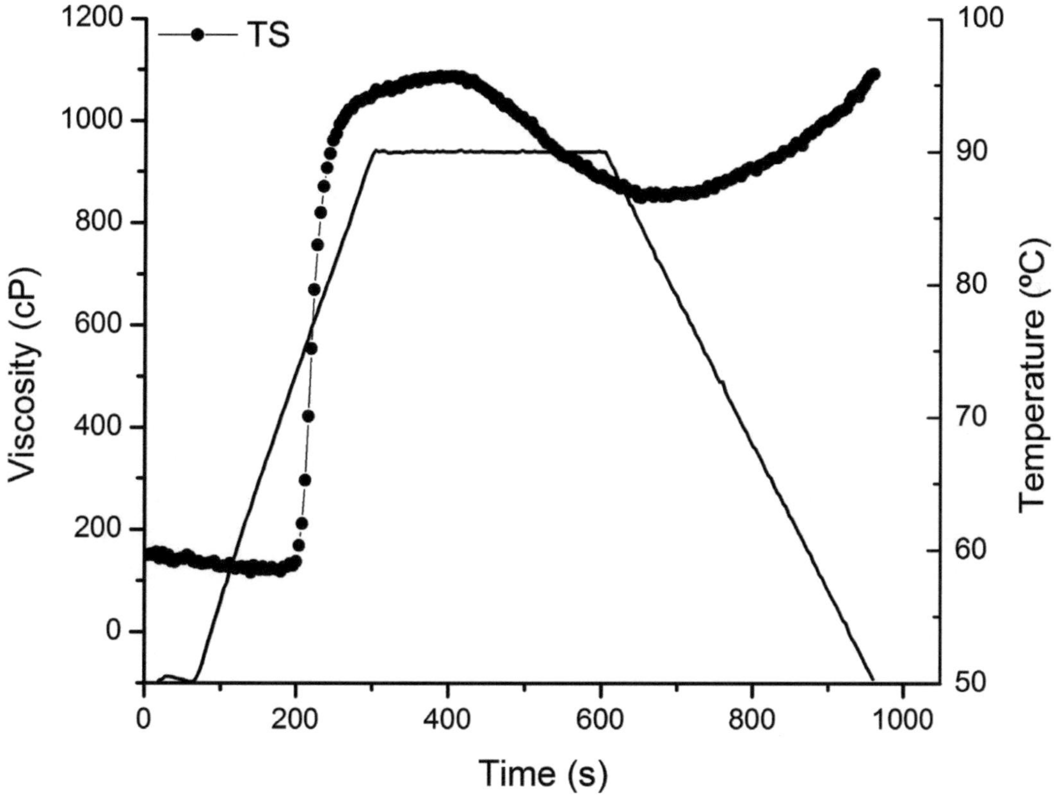

Figure 1. Pasting properties of tapioca starch (TS) (6% w/w).

3.2.2.1. Rheological measurements after pasting: Steady shear measurements

After pasting, the starch molecules tend to re-associate and form a viscous paste; the viscosity of this starch paste can be measured using a viscometer, such as a capillary flow meter, falling ball viscometer, or rotational rheometer.

Generally, the viscosity of starch pastes presents a non-Newtonian feature: the shear stress does not increase linearly with the increase in shear rate. The viscosity of starch paste is also shear thinning (flow behavior index, $n<1$) because they become thinner and decrease viscosity with increasing shear rate and time [8].

At low shear rates, the Brownian motion kept all molecules at random despite the initial effects of shear orientation, and this makes pseudoplastic materials to behave like a Newtonian liquid [25]. At increasing shear rate, the viscosity decrease. This could be explained as follows: during the shear process, the molecules are more or less oriented in shear direction and the shear gradient direction also influences this orientation. The increased motion allows the molecules to disentangle to a certain extent, which reduces their flow resistance. Also, low concentrations of polysaccharides may involve completely disentangled chains with a high degree of orientation. The viscosity is affected by measurement temperature and increases with starch concentration and decreases with amylose content.

For instance, tapioca starch exhibited a wide hysteresis loop. The tapioca starch alone presented more remarkable thixotropy (higher shear stress of upward curve than that of downward curve). The clockwise loop (thixotropy) can be interpreted as a structure breakdown by the shear field to alter a structure or a form a new structure [25]. Waxy corn starch showed a combined hysteresis loop, a clockwise loop, and a counterclockwise loop (antithixotropic behavior), and the prevalence of one behavior against the other is due to the heating conditions. When starch suspensions from waxy maize are pasted completely with more heating time and agitation, the antithixotropic property becomes more significant [9].

3.2.3. Retrogradation

Starch paste usually ascribes to the hot freshly cooked system, and a gel is formed after cooling. The characterization of rheological properties of starch gel has been determined by different methods: using a texture analyzer to determine the starch gel strength as a single point measurement; using a dynamic rheometer that provides the continuous assessment of starch gel at various temperatures and shear rates. The hot pastes begin to cool, after becoming more elastic and start to develop solid properties. Transitions from viscous paste to an elastic gel can be evaluated by storage modulus (G'), and loss modulus (G''), the loss tangent being ($\tan \delta = G''/G'$) [20]. The starch gel can behave more like a solid (a small $\tan \delta$, G' is much larger than G''); this reflects a recoverable deformation or more like a liquid (a large $\tan \delta$, G' is smaller than G''); it means that the energy employed to deform the gel is dissipated. During storage, G' and G'' increase, indicating the strength, and the starch gel becomes firmer [8].

Starch paste developed interactions between amylose and amylopectin molecules, and a formation of networks occurred to hold water in the swollen granules, and a starch gel was formed. The retrogradation is a two-step process; the short-term retrogradation involving primarily amylose gelation developed after cooking and lasts up to 48 h. Native starch with a greater amylose content tends to arrange a stronger gel at a faster rate. Amylose molecules are unstable, and retrogradation results in an increase of turbidity and eventual precipitation. The retrogradation occurs because amylose molecules shrink, which is caused by a decrease in kinetic energy and Brownian motion of the polymer and water molecules. The shrinkage results in the formation of new intra- and inter-molecular hydrogen bonding, and the intensity of these hydrogen bonds leads to precipitation of amylose [26].

The second stage of retrogradation involves the recrystallization of amylopectin, and the process takes a long time due to the presence of short and highly branched side chains (A and B1), which was caused by their high kinetic energy [26].

4. Galactomannans

Galactomannans are heterogeneous polysaccharides found in the endosperm of the seed of a certain variety of legumes formed by a β-(1-4)-D-mannan backbone with a single D-galactose branch linked α-(1-6). They differ from each other by the mannose/galactose (M/G) ratio. The major sources of seed galactomannans are locust bean gum (*Ceratonia siliqua*), with M:G ratio of ~3.5:1, guar gum (*Cyamopsis tetragonoloba*), with M:G ratio of ~2:1) tara (*Caesalpinia spinosa*),with M:G ratio of ~3:1, and fenugreek (*Trigonella foenum-graecum L.*), which has an M:G ratio of ~1:1 [27]. Galactomannans are widely used in the industry due to their functional properties as thickening, binding, and stabilizing agents, and because they are not significantly affected by pH levels, added ions, and/or heating processes [28].

The galactomannans in their basic structure consist of mannose, which contains cis-OH groups in the polymer. Therefore, the ability to form hydrogen bonds between each mannose chain is expected to be long as neighboring groups such as galactose do not develop steric hindrance that prevents each chain of galactomannans from getting too close and prevents the cis-OH groups of the mannose from forming hydrogen bonds. Thus, the galactomannans with additional cross-links via hydrogen bonds are less capable of hydration (locust bean gum, M:G ratio of ~4:1). On the other hand, a greater substitution (guar gum, M:G ratio of ~2:1) leads to improved solubility [27].

Challenges in the food industry bring an interest in starch-galactomannan systems, and researchers have done extensive work in this area. Changes in different stages of starch under heating-cooling processes are reviewed through the addition of different galactomannans.

4.1. Starch-galactomannans system in gelatinization and pasting

It is well known that the addition of hydrocolloids in specific galactomannans affects the rheological properties and influences the increase of viscosity of pastes or solutions. A significant increase of viscosity through the addition of galactomannans has been demonstrated using guar gum, locust bean gum, and fenugreek gum. A soluble complex was evidenced as a result of the molecular associations between the amylose and the polysaccharide, as the main cause for viscosity increases during gelatinization, lowering the pasting temperature

[29]. A decrease in the pasting temperature in the presence of guar gum could be attributed to phase separation, an increasing gum concentration in the continuous phase, and an increased volume fraction of the starch granules (amylopectin) in the disperse phase as a result of the mutual exclusion between leached starch and hydrocolloid molecules based on thermodynamic incompatibility between the two polysaccharides. An increase in the effective concentration of leached starch molecules, primarily amylose and the hydrocolloid, in the continuous phase upon heating enhanced interactions between the swollen starch granules, allowing pasting at a lower temperature. However, interactions between leached starch molecules and hydrocolloid in the continuous phase should not be minimized as a factor that influences the lower pasting temperature [30, 31].

Funami [29] added 0.5% guar gum and fenugreek gum with various molecular weights to starch solutions (15%) and modified the gelatinization behavior of starch. Analyses were done on the RVA. The viscosity profiles and pasting temperature shifted to higher temperatures with increasing M_w of each galactomannan and lower temperatures with starch concentration (5%). When M_w was standardized by the molecular weight of mannose, the effect of each galactomannan on the pasting temperature overlapped each other better than when M_w was standardized by the molecular weight of the galactose, and with a lower starch concentration (5%), pasting temperatures shifted oppositely to lower pasting temperatures with increasing M_w of the polysaccharide.

An increase in the final viscosity and setback values suggested that the retrogradation of starch should be promoted at a very early stage of storage by the addition of galactomannans.

4.2. Starch-galactomannans system after pasting: Flow behavior

The addition of galactomannans also influences the flow behavior of starch pastes. Different authors correlate the effect of galactomannans to changes in flow behavior. Rheograms of rice starch-galactomannan (guar gum and locust bean gum) mixtures show more pseudoplastic behavior of guar gum mixtures (*n* value of 0.17–0.18) than control (0.24) and rice starch-locust bean gum (*n* value, 0.21–0.26) [32]. Other studies in wheat starch-galactomannan mixtures show a similar behavior [33]. These phenomena could be explained as follows: the gum structures of guar gum consist of a mannan backbone with alternate galactose branches presenting much more steric hindrance and little ability to establish hydrogen bonds between links [27]. This keeps the molecule in its extended form, which can readily interact with amylose molecules through noncovalent hydrogen bonding, resulting in an extended conformation that increases the degree of pseudoplasticity. On the other hand, the locust bean gum has a M:G ratio of ~4:1; this lower and irregular degree of substitution of the galactose branches in the mannan backbone explains the galactomannan's structure behavior, which tends to coil with the formation of intramolecular hydrogen bonds and thus interacts less with the linear amylose molecules due to the amount of hydroxyl groups available to form intermolecular hydrogen bonding with amylose are lower [32].

4.2.1. Hysteresis

Figure 2 shows that the addition of guar gum (0.5%) to tapioca starch (6%) reduces the hysteresis loop. The addition of guar gum on different starches such as tapioca, potato, oat, and

http://dx.doi.org/10.5772/intechopen.68915

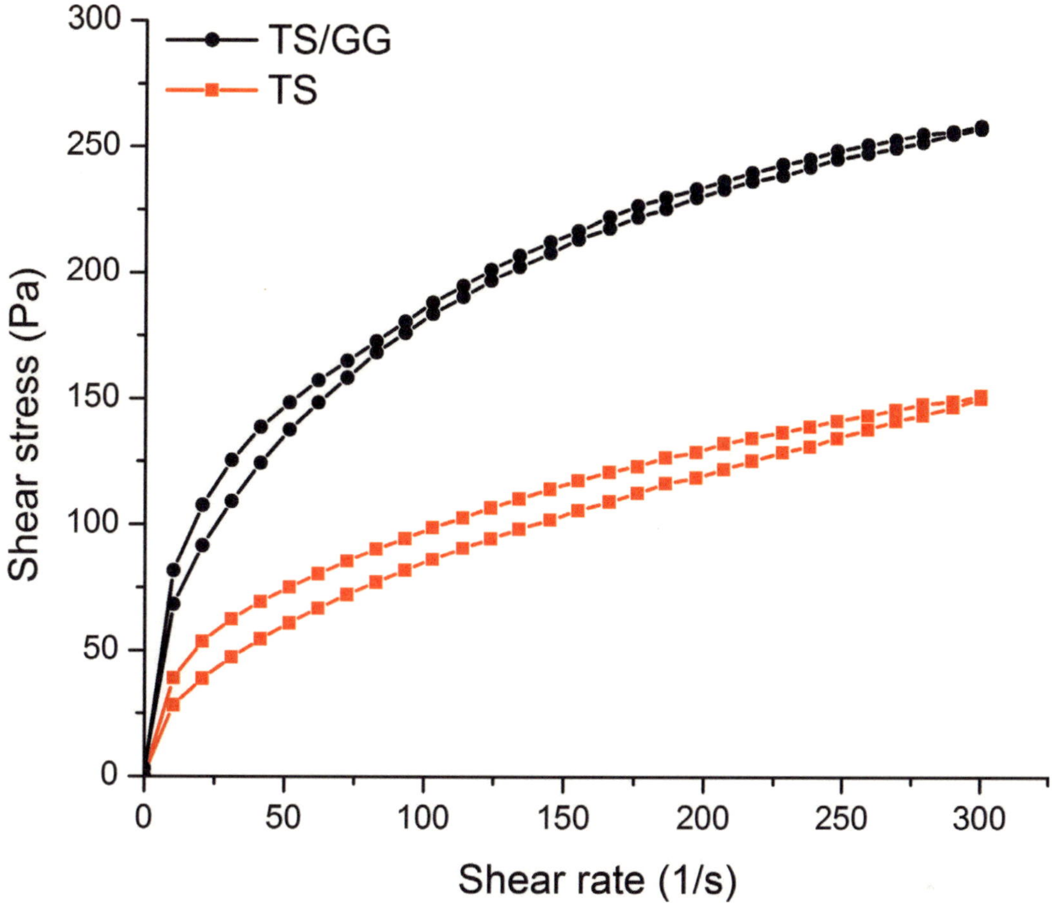

Figure 2. Flow curves of tapioca starch pastes (6% w/w), without guar gum (TS) and 0.5% w/w guar gum (TS/GG).

corn resulted in highly negative hysteresis loops, which meant that in the presence of guar gum, shearing ordered structures generate more readily [34]. Pastes of waxy corn starch with guar gum and waxy corn starch alone exhibited similar combined hysteresis loops, which are anti-clockwise loops (antithixotropic behavior) at low shear rates and clockwise loops (thixotropic behavior) at higher shear rates ($\gamma > 50\ s^{-1}$). These systems contained a large amount of amylopectin (waxy corn starch) that was responsible for shear-thickening properties and shear-induced structure formation and alteration, that is, antithixotropic behavior [9].

4.3. Starch-galactomannans system in retrogradation

Short-term retrogradation is accelerated by the addition of galactomannans in terms of rheology. An increase in the effective concentration of amylose in the continuous phase was suggested as the main cause, leading to the acceleration of retrogradation during storage for shorter periods while preventing molecular ordering or crystallization of starch, particularly the amylopectin fraction of relatively short-chain segments leading to the retardation of retrogradation during storage for longer periods [29]. **Figure 3** shows the variation of G', G'',

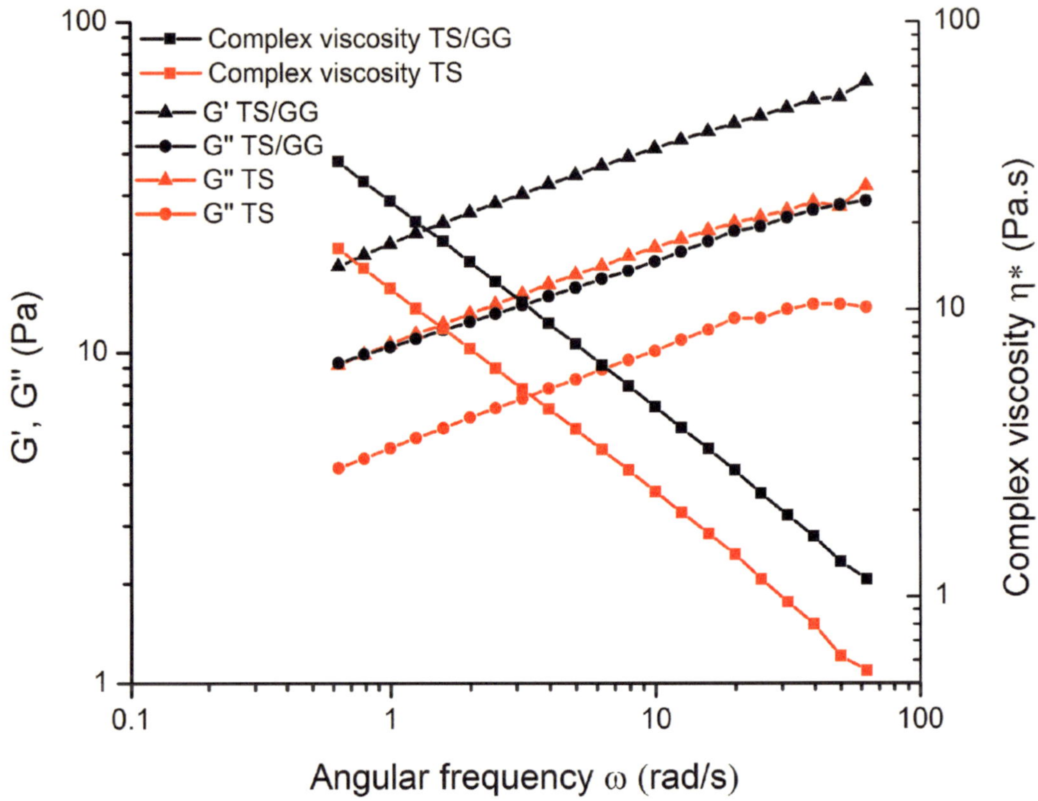

Figure 3. Variation of *G′*, *G″* with angular frequency for 6% w/w TS/guar mixed pastes without guar gum (TS) and 0.5% w/w guar gum (TS/GG) measurements at 25°C and 1.5% strain.

and complex viscosity (η^*) for tapioca starch pastes (6% w/w) alone and with addition of 0.5% (w/w) at 25°C and a strain amplitude of 1.5%. **Figure 4** depicts the variation of tan δ of samples. *G′*, *G″*, and complex viscosity increased with the addition of guar gum. The tan δ values of the starch-guar gum mixture was lower than the tapioca starch alone.

A starch-galactomannans mixture can be classified as a weak gel; in most cases *G′* and *G″* increases with an increase in frequency with a small dependency. Weak gels have intermediate rheological properties between solutions and real gels, under small deformation; weak gels mechanically behave like real gels, but with increasing deformation, the three-dimensional network undergoes a progressive failure into smaller clusters [20], while the tan δ values directly state the *G′*/*G″* ratio.

The tan δ values of rice starch mixed with guar gum and locust bean gum were higher than those on control and increased with an increase in the gum concentration. This means a higher growth of *G″* than *G′* especially on starch-locust bean gum mixtures; the addition of the gums is not able to provide effective contribution to the elastic properties of starch dispersions [35]. On the other hand, a reduction in tan δ by the addition of guar gum on cationic tapioca starch is also presented, similar to what is shown in **Figure 4**, and it is related to a more solid-like

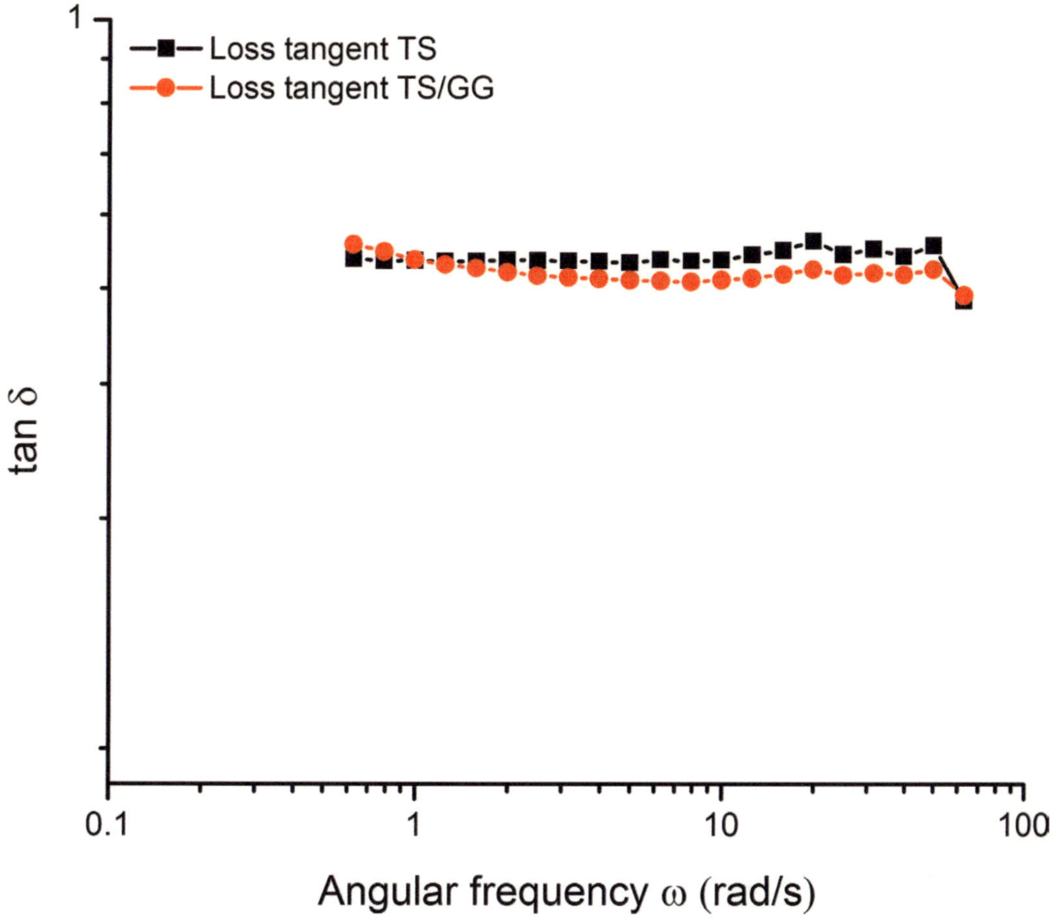

Figure 4. Variation of tan δ with angular frequency for 6% w/w TS/guar mixed pastes without guar gum (TS) and 0.5% w/w guar gum (TS/GG) measurements at 25°C and 1.5% strain.

behavior, that is, a stronger three-dimensional network constructed by amylose and an amylose-gum system [3]. This supports the hypothesis that the change in dynamic rheological properties is due to the thermodynamically incompatible network structure in which interactions between polymers of the same type are favored energetically in comparison with interactions with different polymers [35] and also the interactions of branched or linear molecules of, that is, guar gum that can interact with amylose via noncovalent hydrogen bonding [36].

The addition of these gums during aging (10 h at 4°C) increases the G' value with the gum concentration (0.2–0.8%). The gum structure also had a main role in G' development, while starch-guar gum mixtures exhibit a plateau. The starch-locust bean gum dispersion shows a continuous increase of G'. The unsubstituted smooth regions of mannan backbone in LBG tend to associate with each other, forming a three-dimensional network. The thickening effect of the gums restricts the mobility of amylose, resulting in interactions between neighboring amylose molecules more easily accelerating the short-term retrogradation [35].

The molecular associations between gums and amylopectin fractions avoid the formation of crystalline structures during storage, affecting long-term retrogradation. Corn starch-fenugreek gum mixtures present another factor to inhibit long-term retrogradation: gums can stabilize water molecules; therefore, they can act as a water binder, depriving amylose, or amylopectin of usable water for crystallization [5]. In rice starch-galactomannans systems (guar gum and locust bean gum), it was found that the addition of galactomannans led to an increase in the viscous properties of starch resulting in retardation of retrogradation of starch [35].

5. Applications of starch-galactomannans systems in food development

Extensive research has been carried out on the application of starch-galactomannans systems in food development, formulation of sauces, soups, and caramel sauces, improving the texture, food quality, and stabilization, and in structured foods as functional ingredients that can change the rate of degradation of carbohydrates such as starch during digestion, thus regulating insulin levels, which is key in the prevention of obesity and diabetes.

Food structure has a significant impact on the food digestive behavior [37]. For instance, it has been demonstrated that adding dietary fiber such as galactomannans (guar gum, locust bean gum, and fenugreek gum), pectins, β-glucans, cellulose, and fibers into foods can notably delay the absorption of glucose into the bloodstream [38–40].

Dietary fibers are ingredients of food defined as a heterogeneous mixture of substances predominantly found in plant cell walls that the body cannot digest or absorb. Indigestibility implies that dietary fiber pass without much break-down to the enzymatic action in the upper part of the gastrointestinal tract because enzymes capable of cleaving β-1-4 bonds present in dietary fibers are not found in the human small intestines. Furthermore, dietary fibers or nonstarch polysaccharides (NSP) is used as a "fuel" for the gut microbiota which produces the short-chain fatty acids in the large bowel that are easily absorbed and contribute to energy metabolism [41].

Dietary fiber is classified into two categories: water-soluble/well-fermented fibers and water-insoluble/less-fermented fibers [42]. Soluble viscous fibers contribute to the formation of the unstirred layer in the intestinal lumen, which serves as a physical barrier to fats, including cholesterol and bile acid reabsorption. Water-insoluble fractions such as cellulose and lignin, and water-soluble components comprise the major part of hemicellulose compounds, gums, β-glucans, mucilages, and pectins [43].

The two major benefits associated with soluble and viscous dietary fibers are modulation of glycemic postprandial response and plasma cholesterol lowering [44]. The physiological value of soluble dietary fiber is based on its ability to thicken into swollen hydrated networks and ensue viscosity that shows its potential to exert biological effects through the stomach and small intestines [45]. Studies have shown that the addition of soluble fibers such as pectin, guar gum, and locust bean gum in food reduces gastric emptying, delays absorption of actives, reduces the plasma glucose response, and slows down the return of hunger. Soluble fibers have been combined in diets for treating pathological conditions such as obesity, hypercholesterolemia, and diabetes [46].

5.1. Galactomannans as dietary fibers

Guar gum contains soluble dietary fibers. The guar seed possesses 52–48% total dietary fiber on seed dry weight, and the water-soluble endosperm contributes 26–32% of soluble dietary fiber [47]. Previous studies *in vivo* have found that, when comparing guar gum with pectin and carboxymethilcellulose (CMC) on postprandial glucose and insulin levels, guar showed to be the most effective biopolymer on reducing plasma glucose levels [48]. This behavior was explained by the fact that guar showed the greatest stability and viscosity under acidic conditions *in vitro*, i.e., under gastric conditions. Jaime-Fonseca [41] stated that comparing the effect of guar gum and pectin by measuring the flow behavior and the overall mass transfer coefficient (OMTC) shows a crossover at 100 s^{-1} and values of OMTC very similar.

Locust bean gum has been studied as a galactomannan that falls under the category of viscous soluble fibers, and it was proposed flattening postprandial glycemia more consistently than wheat bran and other insoluble fibers. The addition of locust bean gum has a major impact on gastric emptying by delaying the emptying rate [49].

Fenugreek gum has beneficial physiological effects, including the antidiabetic and hypocholesterolemic effects, attributable to the intrinsic dietary fiber constituent that has promising nutraceutical values [34]. Dietary fiber on seeds is around 45.5% (32% insoluble and 13.3% soluble), which changes the texture of food and has a beneficial influence on digestion [50]. The galactomannan-rich soluble fiber fraction of fenugreek seeds may be responsible for the antidiabetic activity and can slow the rate of postprandial glucose absorption. This may be a secondary mechanism for its hypolglycemic effect [51].

5.2. Addition of galactomannans on starch-based products

5.2.1. Starch-galactomannan aqueous system

The carbohydrate content of food is important due to the postprandial glycemic effect on carbohydrate consumption. The increase in postprandial glycemia with the ingestion of carbohydrates and starch has been shown to be controlled by the addition of galactomannans.

Starch needs to be converted from its native form to the gelatinized state in order to be hydrolyzed and consequently release glucose. Direct cooking by heating to 100°C in excess water is the principal process facilitating starch availability for water penetration and consequent α-amylase action for several minutes [52]. It has been stated that guar gum, an industrial galactomannan, competes with starch for water in food systems and prevents starch gelatinization, thus depriving starch from accessing water [53]. Another hypothesis related to the role of dietary fibers is in the reduction of hydrolysis; therefore, the digestion rate is the capability to form a barrier around starch granules, with the result that starch can resist enzymatic degradation [54, 55].

In order to evaluate the effect of the dietary fibers, the conditions of the gastrointestinal tract (GIT) that comprises the secretion of fluids, enzyme hydrolysis, pH, and the presence of salts are simulated to obtain an *in vitro* digestion model and measure the glucose diffusion. The addition of galactomannans to different starch sources to evaluate whether the viscosity remains after passing through these conditions is reviewed on this section.

For this reason, rheological rotational (flow behavior) and oscillatory (frequency sweep and time sweep) measurements can give valuable information. The rotational tests can be used to mimic the shear rate found during digestion (100 s^{-1}) [11, 56]. It has been reported that shear rates of 10–100 s^{-1} are found during the various processes of digestion. Oscillatory tests, such as time sweep and frequency sweep, can be useful to explain the structural changes during and after digestion, respectively, and demonstrate the rate of degradation of starch/galactomannans entanglements and the solid or liquid-like behavior during these processes. Additionally, viscosity measurements using RVA could be relevant to characterize the behavior of gelatinization-pasting of samples.

Brennan et al. [57] investigated the effect of guar gum, locust bean gum, and Arabic gum on wheat starch and flour viscosity and their effects on *in vitro* starch hydrolysis. Both guar gum and locust bean gum increased the peak and final viscosity on the resulting pastes, whereas Arabic gum significantly reduced the viscosities of pastes. Guar gum and locust bean gum reduced the amount of starch degradation; on the contrary, Arabic gum increased the amount of starch hydrolysis. The researcher thus concluded that the extent of starch hydrolysis was dependent on viscosity changes.

The reduction of starch hydrolysis with different polysaccharides (Arabic, carrageenan, guar, pectin, and xanthan) added to wheat, normal maize starch, and waxy maize starch was found by Tester and Sommerville [53]. The reduction on hydrolysis was related to a reduction in water accessibility and restricting water availability/mobility (due hydration) during the gelatinization event. Otherwise, the starch sources used in the study vary and may be another factor causing differences on glycemic results.

Dartois et al. [7] investigated the effect of guar gum on the digestibility of waxy maize starch *in vitro* under simulated gastric and intestinal conditions. No starch hydrolysis was observed under simulated gastric conditions, whereas more than 90% hydrolysis was observed at the end of digestion-simulated intestinal conditions. The addition of guar gum led to a significant decrease in both the rate and the extent of final starch hydrolysis and showed that starch digestion is not only delayed but also reduced in the presence of gum. Rheological measurements demonstrate that the viscosity of digesta decreased during intestinal digestion, but the extent of decrease was lower in the presence of guar gum. Frequency sweep tests showed that after 5 min of digestion, G' and G'' were higher than samples with 30 s and 1 min of digestion and starch guar gum without enzymatic treatment due to the inhibition to starch components from leaching out of the granules into the continuous phase of starch pastes during gelatinization. Time sweep test demonstrated that G' and G'' increased as digestion progresses and water is liberated, allowing the gum present in the system to take up water.

Fabek et al. [56] evaluated the ability of resist loss viscosity of locust bean gum, guar gum, fenugreek gum, and xanthan gum in a two-stage *in vitro* digestion (gastric and small intestinal) phases and also studied starch (waxy corn starch) and protein (sodium caseinate) addition to simulate a food model for glucose release measurements. Flow behavior measurements showed that dilutions representing watery secretions during digestion were the most important factors to reduce viscosity, and changes in pH have an additive effect to that reduction. All gums demonstrated an ability to depress the glucose response. Xanthan gum had the ability to withstand reductions in viscosity and lowering glucose diffusion, and guar gum was the second; although it has the lowest glucose diffusion ratio among the other gums, it was not significantly different. However, the effect on attenuating glucose diffusion was not as pronounced, suggesting that viscosity alone may not be the only contributing factor in the reduction of postprandial glucose levels.

5.2.2. Starch-galactomannan system in extrusion cooking

Extrusion is a high-temperature, short-time, continuous processing method in which temperature, moisture, pressure, and shearing interact to develop products with functional properties [58]. When starch is processed by extrusion, many phenomena take place, including gelatinization, thermal melting, crystallite loss, and molecular degradation of amylose and amylopectin, which also makes this biopolymer available to digestive enzymes [59]. Low values of digestibility of starch-extruded products have been attributed to the formation of amylose lipid complex, starch protein interaction, and limited water availability that prolongs the starch digestibility during enzymatic hydrolysis. The fact that guar gum competes against starch for available water in gelatinization allows starch hydrolysis to investigate the addition of galactomannans to starch and evaluate its behavior.

Adamu [60] reported that when extruded samples of starch added to guar gum are cooled, re-association of the chains of the polysaccharide and the gum appears. The dissolution of an extruded sample previously added with guar gum in water generates swelling, followed by an interlocking of the chains in the polysaccharide. This provokes an increase in the viscosity of the solution [61].

Brennan et al. [62] evaluated the effects of the addition of dietary fiber (wheat bran, guar gum, inulin, hi-maize, and swede) at different concentrations (5–15%) and found that peak and final viscosities among the different extruded products increased with the addition of guar gum. Guar gum decreased readily digestible starch, being more effective than wheat bran.

Control of extrusion parameters as barrel temperature or moisture content of processing plays a main role in the thermal transitions of starch and its digestibility. Guar gum addition to different flours (maize, potato, rice, and wheat) and processed by extrusion presented low viscosity and has been associated with the gelatinization and disruption of the starch granules during the high shear and high temperature of extrusion. It is also probable that a more extensive starch dextrinization during extrusion of samples containing guar gum results in lesser viscosity of extrudates. These changes on extrudates resulted in starch being more rapidly digestible [63].

It has been proven that extrusion conditions can affect the depolymerization of guar gum; however, guar gum impair increasing viscosity values on corn starch extrudate samples with high moisture contents and low extrusion temperature during processing. Guar gum decreased gelatinization and promoted fusion and depolymerization of starch due to shear forces and temperature during extrusion by competing for the water available on this process and did not affect starch hydrolysis [64].

The effect of adding galactomannans such as guar gum on starch hydrolysis after direct cooking or extrusion cooking is inconclusive, and the mechanisms for the reduction of starch hydrolysis and reduced postprandial glycemia are not fully understood, but increased viscosity might be one of the most important factors. So rheological measurements are proven as an important tool, not only to provide information about texture and quality control, but also for the application of ingredients to elaborate functional foods.

6. Conclusion

In this chapter, rheological techniques were exposed to highlight the transitions of starch in aqueous systems during heating and after processing. The behavior of different starches with galactomannans is presented to understand the rheology of transitions of starch as a measure to control and understand the interactions inside the starch granules and the starch-galactomannan systems that can be used in food products and employed to develop structured foods.

Structural changes that occurred during the heating of starches can be determined using rheological measurements: rotational tests as steady shear measurements to determine the hysteresis loop; oscillatory tests (frequency sweeps, temperature sweeps, and time sweeps) combined with RVA measurements, DSC thermographs, texture analysis, and X-ray diffraction.

Starch modification through the addition of galactomannans is presented based on its interaction with starch molecules. Galactomannans affect the phase separation by occupying the continuous phase and therefore increasing the volume of starch granules in the disperse phase. They compete with starch for the available water, leading to further alterations in the structure and changing the pasting temperature, the gelatinization, and improving viscosities. Pseudoplastic behavior of starch-galactomannan mixtures presents less clockwise loops (thixotropic) and, in some stages, anti-clockwise loops. The less solid-like structure of pastes with galactomannans is shown. The short-term retrogradation is accelerated, and long-term retrogradation is diminished by interfering with amylopectin chains.

The role of rheological properties can be significant to reduce the starch hydrolysis by adding galactomannans to starch food products due to the increased viscosity of the system, which may have an affect not only on the mass transfer (diffusion coefficient) of the molecules but also on enzymatic reactions.

Novel applications and new galactomannans employments will fulfill the starch-based products field. Furthermore, the research and interests of finding new ways to improve food structure is always a never-ending discovery.

Author details

Erich von Borries-Medrano*, Mónica R. Jaime-Fonseca and Miguel Á. Aguilar-Méndez

*Address all correspondence to: evborries@hotmail.com

Instituto Politécnico Nacional, Centro de Investigación en Ciencia Aplicada y Tecnología Avanzada, Unidad Legaria, Ciudad de México, México

References

[1] Wang S, Li C, Copeland L, Niu Q, Wang S. Starch retrogradation: A comprehensive review. Comprehensive Reviews in Food Science and Food Safety. 2015;**14**:568-585

[2] Qiu S, Yadav MP, Chen H, Liu Y, Tatsumi E, Yin L. Effects of corn fiber gum (CFG) on the pasting and thermal behaviors of maize starch. Carbohydrate Polymers. 2015;**115**:246-252

[3] Chaisawang M, Suphantharika M. Effects of guar gum and xanthan gum additions on physical and rheological properties of cationic tapioca starch. Carbohydrate Polymers. 2005;**61**:288-295

[4] Achayuthakan P, Suphantharika M. Pasting and rheological properties of waxy corn starch as affected by guar gum and xanthan gum. Carbohydrate Polymers. 2008;**71**:9-17

[5] Funami T, Kataoka Y, Noda S, Hiroe M, Ishihara S, Asai I, et al. Functions of fenugreek gum with various molecular weights on the gelatinization and retrogradation behaviors of corn starch—1: Characterizations of fenugreek gum and investigations of corn starch/fenugreek gum composite system at a relatively high starch concentration; 15w/v%. Food Hydrocolloids. 2008;**22**:763-776

[6] Funami T, Kataoka Y, Omoto T, Goto Y, Asai I, Nishinari K.Food hydrocolloids control the gelatinization and retrogradation behavior of starch. 2a. Functions of guar gums with different molecular weights on the gelatinization behavior of corn starch. Food Hydrocolloids. 2005;**19**:15-24

[7] Dartois A, Singh J, Kaur L, Singh H. Influence of guar gum on the in vitro starch digestibility—Rheological and microstructural characteristics. Food Biophysics. 2010;**5**:149-160

[8] Ai Y, Jane JI. Gelatinization and rheological properties of starch. Starch-Stärke. 2015;**67**: 213-224

[9] Wang B, Wang L-J, Li D, Özkan N, Li S-J, Mao Z-H. Rheological properties of waxy maize starch and xanthan gum mixtures in the presence of sucrose. Carbohydrate Polymers. 2009;**77**:472-481

[10] Sikora M, Krystyjan M, Tomasik P, Krawontka J. Mixed pastes of starches with guar gum. Polimery. 2010;**55**:582-590

[11] Steffe JF. Rheological methods in food process engineering. East Lansing, MI, USA: Freeman press; 1996.

[12] Özkan N, Xin H, Chen XD. Application of a depth sensing indentation hardness test to evaluate the mechanical properties of food materials. Journal of Food Science. 2002;**67**:1814-1820

[13] Amini AM, Razavi SMA, Mortazavi SA. Morphological, physicochemical, and viscoelastic properties of sonicated corn starch. Carbohydrate Polymers. 2015;**122**:282-292

[14] Robyt JF. Starch: Structure, properties, chemistry, and enzymology. In: Fraser-Reid BO, Tatsuta K, Thiem J, editors. Glycoscience: Chemistry and Chemical Biology. Berlin, Heidelberg: Springer Berlin Heidelberg; 2008. p. 1437-1472.

[15] Xie F, Halley PJ, Avérous L. Rheology to understand and optimize processibility, structures and properties of starch polymeric materials. Progress in Polymer Science. 2012;**37**:595-623

[16] Colonna P, Buleon A. Thermal transitions of starches. In: Bertolini AC, editor. Starches: Characterization, Properties, and Applications. 1st ed. Boca Raton, FL: CRC Press; 2010. pp. 71-102

[17] Ahmed J, Ramaswamy HS, Ayad A, Alli I. Thermal and dynamic rheology of insoluble starch from basmati rice. Food Hydrocolloids. 2008;**22**:278-287

[18] Liu C-M, Liang R-H, Dai T-T, Ye J-P, Zeng Z-C, Luo S-J, et al. Effect of dynamic high pressure microfluidization modified insoluble dietary fiber on gelatinization and rheology of rice starch. Food Hydrocolloids. 2016;**57**:55-61

[19] Singh N, Singh J, Kaur L, Singh Sodhi N, Singh GB. Morphological, thermal and rheological properties of starches from different botanical sources. Food Chemistry. 2003;**81**:219-231

[20] Yousefi A, Razavi S. Dynamic rheological properties of wheat starch gels as affected by chemical modification and concentration. Starch-Stärke. 2015;**67**:567-576

[21] Jane J, Chen Y, Lee L, McPherson A, Wong K, Radosavljevic M, et al. Effects of amylopectin branch chain length and amylose content on the gelatinization and pasting properties of starch 1. Cereal Chemistry. 1999;**76**:629-637

[22] Singh J, Singh N. Studies on the morphological, thermal and rheological properties of starch separated from some Indian potato cultivars. Food Chemistry. 2001;**75**:67-77

[23] Singh N, Singh J, Kaur L, Sodhi NS, Gill BS. Morphological, thermal and rheological properties of starches from different botanical sources. Food Chemistry. 2003;**81**:219-231

[24] Lii C-Y, Tsai M-L, Tseng K-H. Effect of amylose content on the rheological property of rice starch. Cereal chemistry. 1996;**73**:415-420

[25] Hong Y, Zhu L, Gu Z. Effects of sugar, salt and acid on tapioca starch and tapioca starch-xanthan gum combinations. Starch-Stärke. 2014;**66**:436-443

[26] Tako M, Tamaki Y, Teruya T, Takeda Y.The Principles of Starch Gelatinization and Retrogradation. Food and Nutrition. 2014;**5**:280-291.

[27] Prajapati VD, Jani GK, Moradiya NG, Randeria NP, Nagar BJ, Naikwadi NN, et al. Galactomannan: A versatile biodegradable seed polysaccharide. International Journal of Biological Macromolecules. 2013;**60**:83-92

[28] Sittikijyothin W, Torres D, Gonçalves M. Modelling the rheological behaviour of galactomannan aqueous solutions. Carbohydrate Polymers. 2005;**59**:339-350

[29] Funami T. Functions of food polysaccharides to control the gelatinization and retrogradation behaviors of starch in an aqueous system in relation to the macromolecular characteristics of food polysaccharides. Food Science and Technology Research. 2009;**15**:557-568

[30] Alloncle M, Lefebvre J, Llamas G, Doublier J. A rheological characterization of cereal starch-galactomannan mixtures. Cereal Chemistry. 1989;**66**:90-93

[31] Samutsri W, Suphantharika M. Effect of salts on pasting, thermal, and rheological properties of rice starch in the presence of non-ionic and ionic hydrocolloids. Carbohydrate Polymers. 2012;**87**:1559-1568

[32] Yoo D, Kim C, Yoo B. Steady and dynamic shear rheology of rice starch-galactomannan mixtures. Starch-Stärke. 2005;**57**:310-318

[33] Sajjan SU, Rao MR. Effect of hydrocolloids on the rheological properties of wheat starch. Carbohydrate Polymers. 1987;**7**:395-402

[34] Srinivasan K. Fenugreek (*Trigonella foenum-graecum*): A review of health beneficial physiological effects. Food Reviews International. 2006;**22**:203-224

[35] Kim C, Lee SP, Yoo B. Dynamic rheology of rice starch-galactomannan mixtures in the aging process. Starch-Stärke. 2006;**58**:35-43

[36] Sudhakar V, Singhal R, Kulkarni P.Starch-galactomannan interactions: Functionality and rheological aspects. Food Chemistry. 1996;**55**:259-264

[37] Lundin L, Golding M, Wooster TJ. Understanding food structure and function in developing food for appetite control. Nutrition & Dietetics. 2008;**65**:S79-S85

[38] Blackburn N, Holgate A, Read N. Does guar gum improve post-prandial hyperglycaemia in humans by reducing small intestinal contact area? British Journal of Nutrition. 1984;**52**:197-204

[39] Jenkins D, Wolever T, Leeds AR, Gassull MA, Haisman P, Dilawari J, et al. Dietary fibres, fibre analogues, and glucose tolerance: importance of viscosity. British Medical Journal. 1978;**1**:1392-1394

[40] Salmerón J, Ascherio A, Rimm EB, Colditz GA, Spiegelman D, Jenkins DJ, et al. Dietary fiber, glycemic load, and risk of NIDDM in men. Diabetes Care. 1997;**20**:545-550

[41] Jaime-Fonseca MR. An Engineering Understanding of the Small Intestine. University of Birmingham; 2012

[42] Dhingra D, Michael M, Rajput H, Patil R. Dietary fibre in foods: A review. Journal of food Science and Technology. 2012;**49**:255-266

[43] Ryttig K. Dietary fibre and weight control In: Mela DJ, editor. Food, Diet and Obesity. 1st ed. Boca Raton, USA: Woodhead Publishing; 2005. p. 311-328.

[44] Anttila H, Sontag-Strohm T, Salovaara H. Viscosity of beta-glucan in oat products. Agricultural and Food Science. 2004;**13**:80-87

[45] Brownlee IA. The physiological roles of dietary fibre. Food Hydrocolloids. 2011;**25**:238-250

[46] Kong F, Singh R.Disintegration of solid foods in human stomach. Journal of food science. 2008;**73**:67-80.

[47] Mudgil D, Barak S, Khatkar BS.Guar gum: processing, properties and food applications—a review. Journal of food Science and technology. 2014;**51**:409-418.

[48] Brenelli S, Campos S, Saad M. Viscosity of gums in vitro and their ability to reduce postprandial hyperglycemia in normal subjects. Brazilian Journal of Medical and Biological Research. 1997;**30**:1437-1440

[49] Barak S, Mudgil D. Locust bean gum: Processing, properties and food applications—A review. International Journal of Biological Macromolecules. 2014;**66**:74-80

[50] Wani SA, Kumar P. Fenugreek: A review on its nutraceutical properties and utilization in various food products. Journal of the Saudi Society of Agricultural Sciences. 2016

[51] Basch E, Ulbricht C, Kuo G, Szapary P, Smith M.Therapeutic applications of fenugreek. Alternative Medicine Review. 2003;**8**:20-27

[52] Singh J, Dartois A, Kaur L.Starch digestibility in food matrix: A review. Trends in Food Science & Technology. 2010;**21**:168-180

[53] Tester R, Sommerville M.The effects of non-starch polysaccharides on the extent of gelatinisation, swelling and α-amylase hydrolysis of maize and wheat starches. Food Hydrocolloids. 2003;**17**:41-54

[54] Bordoloi A, Singh J, Kaur L. In vitro digestibility of starch in cooked potatoes as affected by guar gum: Microstructural and rheological characteristics. Food Chemistry. 2012;**133**:1206-1213

[55] Aravind N, Sissons M, Fellows CM.Effect of soluble fibre (guar gum and carboxymethylcellulose) addition on technological, sensory and structural properties of durum wheat spaghetti. Food Chemistry. 2012;**131**:893-900

[56] Fabek H, Messerschmidt S, Brulport V, Goff HD. The effect of in vitro digestive processes on the viscosity of dietary fibres and their influence on glucose diffusion. Food Hydrocolloids. 2014;**35**:718-726

[57] Brennan CS, Suter M, Luethi T, Matia-Merino L, Qvortrup J. The relationship between wheat flour and starch pasting properties and starch hydrolysis: Effect of non-starch polysaccharides in a starch gel system. Starch-Stärke. 2008;**60**:23-33

[58] Castells M, Marin S, Sanchis V, Ramos A. Fate of mycotoxins in cereals during extrusion cooking: A review. Food Additives and Contaminants. 2005;**22**:150-157

[59] Chiu C-W, Solarek D. Modification of starches. In: Whistler RL, Bemiller JN, editor. Starch Chemistry and Technology. 3rd ed. Orlando, FL: Academic Press; 2009. pp. 629-655.

[60] Adamu B. Resistant starch derived from extruded corn starch and guar gum as affected by acid and surfactants: Structural characterization. Starch-Stärke. 2001;**53**:582-591

[61] Fox JE. Seed gums. In: Imeson A, editor. Thickening and Gelling Agents for Food. London, UK: Blackie Academic & Professional; 1992. pp. 153-170

[62] Brennan MA, Merts I, Monro J, Woolnough J, Brennan CS. Impact of guar and wheat bran on the physical and nutritional quality of extruded breakfast cereals. Starch-Stärke. 2008;**60**:248-256

[63] Parada J, Aguilera JM, Brennan C. Effect of guar gum content on some physical and nutritional properties of extruded products. Journal of Food Engineering. 2011;**103**:324-332

[64] von Borries-Medrano E, Jaime-Fonseca MR, Aguilar-Méndez MA. Starch–guar gum extrudates: Microstructure, physicochemical properties and in-vitro digestion. Food Chemistry. 2016;**194**:891-899

Chapter 6

Antioxidant Activity of Sulfated Seaweeds Polysaccharides by Novel Assisted Extraction

Shao-Chi Wu

Additional information is available at the end of the chapter

http://dx.doi.org/10.5772/intechopen.69633

Abstract

Seaweeds have an extremely numerous of species in the world and been able to be divided into several developmental systems. Broadly, three types of seaweeds can be defined according to their color: brown seaweeds, green seaweeds, and red seaweeds. Thousands of years ago, mankind used seaweeds as food and medicine. Seaweed extracts are gaining increasing attention due to their unique composition and the potential for widespread use in industry. A variety of novel (green) extraction techniques have been devised for converting seaweed biomass into seaweed extracts, such as enzyme-assisted extraction (EAE), microwave-assisted extraction (MAE), pressurized liquid extraction (PLE), supercritical fluid extraction (SFE), and ultrasound-assisted extraction (UAE), which are capable of extracting seaweeds' biologically active compounds without causing degradation. Seaweed extracts contain compounds, such as carbohydrates, proteins, minerals, oils, fats, and polyunsaturated fatty acids and abundant bioactive compounds, such as antioxidants, pigments, and sulfated seaweed polysaccharides (SWP), as well as antibacterial, antifungal, anti-inflammatory, antioxidation, antitumor, antiviral, and hypolipidemic effect. The purpose of this article is to describe the antioxidant activity of SWP of brown seaweeds, green seaweeds, and red seaweeds by novel assisted extraction.

Keywords: antioxidant activity, assisted extraction, sulfated seaweed polysaccharides

1. Introduction

Edible seaweeds are a good source of antioxidants, dietary fibers, essential amino acids, vitamins, phytochemicals, polyunsaturated fatty acids, and minerals [1]. Most seaweed polysaccharides (SWP) are derived from natural sources, such as agarose, alginates, carrageenan, fucoidan, porphyran, and ulvan. Seaweed polysaccharide is a very important biological macromolecule in marine life. Seaweed polysaccharides exhibit a wide range of structure and are still underexploited and should therefore be considered as a novel source of natural products for pharmaceutical discovery [2].

Seaweeds are plant-like organisms that generally live attached to rock or other hard substrata in coastal areas. They can be classified into three different groups, empirically distinguished since the mid-nineteenth century on the basis of thallus color: brown seaweeds (Phaeophyceae), green seaweeds (Chlorophyceae), and red seaweeds (Rhodophyceae) [3, 4]. Among marine resources, seaweeds, which are sometimes known as algae, are well-known natural sources of polysaccharides. SWP are well-known natural seaweed sources of polysaccharides which have considerable numerous bioactive compounds with significant biological features. SWP are most common in the seaweed cell walls, and their number and chemical structure are varying according to the specific seaweeds' species [5]. Bioactive SWP extracted from seaweeds can be classified into three types. The major SWP found in brown seaweeds were fucan and fucoidan; green seaweeds were sulfated rhamnans and ulvan; red seaweeds were galactan and carrageenan [6]. It displays several physicochemical and biological features of potential interest for agricultural, chemical, food, and pharmaceutical applications [7].

The purpose of this chapter is to understand the potential applications of SWP in antioxidant activity of brown seaweeds, green seaweeds, and red seaweeds by enzyme-assisted extraction (EAE), microwave-assisted extraction (MAE), pressurized liquid extraction (PLE), supercritical fluid extraction (SFE), and ultrasound-assisted extraction (UAE), also named green-assisted extraction in recent literary.

2. Novel assisted extraction

Marine plant materials are increasingly being used to isolate and purify bioactive compounds; and then recent studies have reported on the antioxidant potential of SWP of seaweeds [8, 9]. Traditional techniques involve application of solid-liquid extraction (SLE) simply by means of solvent application and leaching. A domestic application of conventional solvent extraction (CSE) is quite familiar to everybody in daily life from making of coffee or tea at home. SLE encompasses conventional methods: Soxhlet extraction (SE), percolation, and maceration extraction (ME). These techniques have been utilized for more than a century for the separation of SWP. However, certain disadvantages pertaining to CSE render its application quite uneconomically due to excessive consumption of energy, polluting solvents, and time. These underlying drawbacks have triggered research that explores more cost-effective and greener techniques for the extraction of SWP from a wide range of seaweed matrices [10]. They include EAE, MAE, PLE, SFE, and UAE techniques [11]. The advantages and disadvantages of these novel assisted extraction methods are shown in **Table 1**.

2.1. EAE

Enzymes can be derived from animal organs, bacteria, fruit extracts, fungi, or vegetable. All these known enzymes are classified according to six basic groups. These categories are organized according to how the enzyme works on a molecular level. These six types of

Novel assisted extraction methods	Advantages	Disadvantages	References
EAE	1. Non-consumed during reaction 2. High conversion yield 3. Nontoxic and biodegradable 4. High selectivity 5. High specificity 6. Utilization in soft conditions 7. Large-scale production	1. Stability 2. Cost	[23]
MAE	1. Decreased in extraction time 2. To avoid the loss of volatile substances during microwave 3. Less solvent is required because no evaporation occurs 4. No hazardous fumes during acid microwave since it is a closed vessel	1. High pressure used poses safety risks 2. The usual constituent material of the vessel does not allow high solution temperatures 3. Addition of reagents is impossible sine it is a single-step procedure 4. Vessel must be cooled down before it can be opened to prevent loss of volatile constituents	[24–26]
PLE	1. Better for increased operating temperature 2. Increase selectivity 3. Precision and reproducibility 4. Reduces oxidation risk 5. Relatively simple compared to SFE 6. Shorter extraction time 7. Reduced solvent consumption 8. Possibility for automation	1. Thermal degradation for thermolabile compounds is a cause for concern 2. Selectivity is mainly affected by varying the solvent type 3. Post-extraction cleanup step is still necessary	[27]
SFE	1. Enhanced extraction efficiency 2. Tunability of the solvent strength 3. Low organic solvent consumption 4. Preservation of bioactive properties 5. Organoleptic properties of the extracts 6. Inline integration with sample preparation and detection methods	1. High-capital investment 2. Large number of variables to optimize 3. Strong dependence on matrix analyte interactions 4. Difficulties in scale-up 5. Difficulties in technology transfer 6. Difficulty in implementing continuous extraction processes 7. Difficulty of extracting more polar compounds	[27]
UAE	1. Easy to use 2. Short time of extraction 3. Small amount of solvent	1. No good recoveries for most PCB congeners 2. Expensive system	[28]

Table 1. Advantages and disadvantages of novel assisted extractions.

enzymes are as follows: hydrolases, isomerases, ligases, lyases, oxidoreductases, and transferases. Hydrolases are the most common type, followed by oxidoreductases and transferases. Enzymes are ideal catalysts that can assist in the extraction of complex bioactive compounds of natural origin by degrading the plant cell walls and membranes. Consequently, they increase plant cell wall permeability, and thus higher extraction yields of bioactive compounds are achieved [12].

Seaweed cell walls are composed of a diverse array of fibrillar, matrix, and crystalline polymers, that is, sulfated and branched polysaccharides, interacting with proteins, various ions, and water. It is necessary to break down seaweed cell walls with enzymes that can be used to remove the cell wall specifically under optimal temperature and pH conditions and then get the desired bioactive compounds [11], depending on what organism you work with, which can be agarase [6], Alcalase [13], carragenanase [14], Celluclast [15], Kojizyme [16], Neutrase [17], Termamyl [18], Ultraflo [19], Umamizyme [20], Viscozyme [21], and xylanase [22].

2.2. MAE

MAE is a novel technique that has many advantages including high extraction efficiency, low solvent consumption, high-purity extracts, and shortened extraction time, which make it well suited for the extraction of bioactive compounds from plant materials [29]. The mechanism of microwave volumetric heating involves the inherent ionic conduction and dipolar relaxation inside a dielectric material. Microwave irradiation induces the rapid elevating temperature of solvent to accelerate the diffusion of pure solvent into plant matrix, as well as the dissolution of the targeted compound into solvent [30]. Microwave energy penetration causes quick elevation of temperature to build the internal pressure inside the cell of plant material. The high interior pressure may destroy the cell wall of plant material to easily release bioactive compounds into solvents [26]. High temperature would cause the dehydration of cellulose and reduce its mechanical strength in MAE, which promotes the solvent to penetrate into the cellular channels and subsequently increase the extraction yield [31].

Microwave is an electromagnetic radiation with wavelengths ranging from 1 m to 1 mm, with frequencies between 300 MHz (100 cm) and 300 GHz (0.1 cm), which can be transmitted as the wave [32]. When microwave passes through the seaweed medium, its energy may be absorbed and converted into thermal energy. Heating using microwave energy is based on two principles: (1) ionic conduction refers to the electrophoretic migration of the charge carriers (e.g. ions and electrons) under the influence of the electric field produced by microwave [33] and (2) dipole rotation happens when the dipolar molecules attempt to follow the electric field in the same alignment [34].

There are two main types of MAE systems available for industrial and commercial applications for natural product extraction: (1) the closed-vessel system and (2) the open-vessel system [10]. In the closed-vessel system, extraction is carried out under controlled conditions of temperature and pressure. This is generally employed for extractions under extremely high-temperature conditions. Diffused microwaves from a cavity magnetron radiate in all directions to interact with plant samples placed in extraction vessels in a closed-vessel chamber. Owing to the even dispersion of microwaves, this technique is also known as the multimode system [35]. In the open-vessel system, also known as the monomode system, the extraction vessel is partially exposed to microwave radiation (focused radiation). A circular metallic waveguide directs the focused microwaves toward the extraction vessel inside the microwave (monomode) cavity. This interaction promotes the initiation of mass transfer between the solute and extractant upon solvation [36].

2.3. PLE

PLE is another novel assisted extraction technique for natural product extraction from food and botanical sample matrices. The Dionex Corporation was the first to introduce the PLE technique as an accelerated solvent extraction (ASE) technology (ASER®) in 1995 [37]. PLE is called accelerated solvent extraction (ASE), enhanced solvent extraction (ESE), pressurized fluid extraction (PFE), or pressurized solvent extraction (PSE). When the solvent used is water, it is common to use other terms, such as high-temperature water extraction (HTWE), hot liquid water extraction (HLWE), hot water extraction (HWE), pressurized hot water extraction (PHWE), subcritical water extraction (SWE), or superheated water extraction (SWE) [38].

The principle of the PLE is based on using elevated temperatures (50–200°C) and pressures (50–150 atm) to extract analytes from solid or semisolid samples within short periods of time (5–15 min). PLE is similar to Soxhlet extraction, except that during the extraction process, the solvent condition inside the PLE cell approaches the supercritical region which results in more efficient extractions. Depending on the temperature at which the extraction is performed, PLE allows the sample to become more soluble and achieve a higher diffusion rate, while the elevated pressure keeps the solvent below its boiling point [39]. PLE permits high extraction efficiency with a low solvent volume (15–40 mL) and a short extraction time (15–20 min). PLE has used less solvent in a shorter period of time, and in oxygen and light-free environment, it has the potential to be a powerful tool in industry [40–42].

2.4. SFE

SFE is the process of separating one component (the extractant) from another (the matrix) using supercritical fluids as the extracting solvent [43]. The supercritical fluid state occurs when a fluid is above its critical temperature (Tc) and critical pressure (Pc), when it is between the typical gas and liquid state. Manipulating the temperature and pressure of the fluid can solubilize the material of interest and selectively extract it. The unique physical properties of supercritical fluids have values for density, diffusivity, and viscosity between liquids and gases. Moreover, near-zero surface tension as well as low viscosities similar to gases allow supercritical fluids to easily penetrate into a microporous matrix material to extract desired compounds [44].

Carbon dioxide (CO_2) is the most used supercritical fluid, sometimes modified by cosolvents, such as ethanol or methanol. Since the critical temperature and critical pressure of CO_2 are only 31°C and 74 bar, extraction is done at temperatures that will not damage heat labile molecules, and the absence of oxygen minimizes oxidation. Thus, an extraction process can take 10–60 min with SFE, while it would take hours or even days with classical methods [45, 46]. On the other hand, it is to be noticed that no organic residue is found both in extract and solid residue and no thermal degradation appears, which results in very high-quality products [47].

2.5. UAE

Ultrasound is sound waves with frequencies higher than the upper audible limit of human hearing. This limit varies from person to person and is approximately 20 kHz (20,000 Hz) in healthy young adults. Ultrasound devices operate with frequencies from 20 kHz up to several GHz [48]. Sound waves produced by an ultrasonic probe allow greater penetration of solvent into the seaweeds, and ultrasonic power also produces high-energy cavitation bubbles containing solvent vapor. The bubbles implode near seaweed walls causing very high local temperatures, pressure increase, and seaweed cell wall destruction, which eases mass transfer from cell to solvent and enhances microstreaming [49]. This effect is much stronger at low frequencies (18–40 kHz) [50].

UAE has emerged as a promising technique that fulfills the required criteria as an inexpensive green extraction technique providing higher recovery of targeted compounds with lower solvent consumption and/or faster analysis and bioactivity properties. Notable UAE features include cost-effectiveness, eco-friendliness, rapidity, simplicity, safety, and versatility, due to the reduced consumption of time, energy, and expensive organic solvents, which is in contrast to traditional extraction techniques [10].

3. Sulfated seaweed polysaccharides (SWP)

Seaweeds are the most abundant source of nonanimal in nature. SWP from some seaweeds have become very important products in the food industry and also possess biological activity of potential medicinal value, such as anti-allergy, anticancer, anticoagulant, anti-inflammation, antioxidant, and antiviral [9, 51–53]. SWP are commonly found in three major groups of seaweeds: brown seaweeds (Phaeophyta), green seaweeds (Chlorophyta), and red seaweeds (Rhodophyta). The major SWP of brown seaweeds are fucans, including fucoidan, sargassan, ascophyllan, and glucuronoxylofucan; and those of red seaweeds are galactans commercially known as agar and carrageenan. On the other hand, the major SWP of green seaweeds are usually sulfated heteropolysaccharides that contain galactose, xylose, arabinose, mannose, glucuronic acid, or glucose [54].

3.1. Brown seaweeds

There are about 1800 species of brown algae, and most are marine. In general, brown algae are larger and more species are found in colder waters. Brown seaweeds are usually grown or collected for food consumption and especially known for their health benefits and high nutritional value, such as kombu or haidai (*Laminaria japonica*), wakame or quandai-cai (*Undaria pinnatifida*), hijiki (*Hizikia fusiforme*), and mozuku (*Cladosiphon okamuranus*). The major fucoidan yielding brown seaweed genera are *Fucus, Sargassum, Laminaria,* and *Undaria* [55].

The term fucoidan is commonly applied for complex SWP, often isolated from seaweeds, mainly containing fucose residues but also many other monosaccharides [7]. Furthermore, fucoidan has a backbone built of (1→3)-linked α-l-fucopyranosyl or of alternating (1→3)- and (1→4)-linked α-l-fucopyranosyl residues, also including sulfated galactofucans with backbones built of (1→6)-β-ᴅ-galacto- and/or (1→2)-β-d-mannopyranosyl units with fucose or fuco-oligosaccharide branching and/or glucuronic acid, xylose, or glucose substitutions. There are at least two distinct forms of fucoidan: U-fucoidan, which is approximately 20% glucuronic acid, and F-fucoidan, which is >95% composed of sulfated esters of fucose [56].

Fucoidans with greater molecular masses and higher degrees of sulfatation form solutions of higher viscosity. Adding glycerol and diols also leads to a significant increase in viscosity [57]. Rheological characteristics of fucoidan from *C. okamuranus* showed shear thinning behavior below 1.5% (W/V) but plastic behavior at 2.0% (W/V). The dynamic viscoelasticity of the fucoidan increased linearly with an increase in concentration and decreased gradually with increase in temperature [58]. Fucoidan with covalent linkage of bovine serum albumin had emulsifying properties, high solubility after heating, and high melting temperature [59]. Crude fucoidan from *Sargassum* sp. demonstrated good emulsion-stabilizing capacities, especially with cedar wood oil and xylene [60].

3.2. Green seaweeds

Green seaweeds are usually grown or collected for food consumption and especially known for their health benefits and high nutritional value, such as aonori, hirohano-hitoegusa nori, or hitoegusa-nori (*Monostroma* spp.) or green laver (*Enteromorpha* spp.). SWP from green seaweeds can be found in *Caulerpa* (sulfated galacotans), *Codium* (sulfated arabinogalactans), *Enteromorpha* (ulvans), *Monostroma* (sulfated rhamnans), and *Ulva* (ulvans) [61].

The structural diversity of SWP found in seaweeds varies with species. Water-soluble extract polysaccharides from *Caulerpa* are mainly composed of glucans and SWP. Heteropolysaccharides (SWP) from *Caulerpa* consist of different monosaccharides, such as galactose, glucose, mannose, and xylose. Among them, galactose is the major sugar source, while glucose, mannose, and xylose are common components. The water-soluble fraction obtained from *Caulerpa sertularioides* with antimicrobial effects which grown under natural conditions contains sulfated galactans constituted of (1→3)-β-D-Gal and (1→6)-β-D-Gal units, and sulfation is observed to occur at the C-2 position of the residues [62]. A water extraction of SWP from *Codium divaricatum* with anticoagulant activity is a galactan which is highly sulfated and substituted with pyruvic acid ketals was mainly constituted of (1→3)-β-D-galactopyranose residues, branched by single (1→)-β-D-galactopyranose units, and the backbone of CP2-1 attached to the main chain at C-4 positions [63]. *Monostroma nitidum* extracted with boiling water could obtain rhamnan sulfate with antithrombin active that consists of α-1,3-linked L-rhamnose residues, some substituted with sulfate groups mainly at position O-2. Minor amounts were also exist internal 1,2-linked rhamnose and branched rhamnose linkages [64]. The structure of an ulvan with anticancer obtained by water extraction from *Ulva lactuca* consists of galactose, glucose, mannose, rhamnose, xylose, uronic acid, and sulfate content. This ulvan is mainly composed of disaccharide [→4)-β-d-GlcA-(1→4)-α-l-Rha3S-(1→] and other minor disaccharides β-GlcA-(1→2)-α-Xyl and β-GlcA-(→2)-α-Rha [65].

Dielectric properties of aqueous solutions from ulvan (*U. meridionalis*) and rhamnan sulfate (*M. lattisimum*) with H^+-form hydrocolloids possess significant improvement in hydration function [66]. Ulvan of *Ulva armoricana* and *Ulva rotundata* showed that chemical structure, macromolecular characteristics, and rheological properties were affected by both species and seasons. The proportion of high-molecular-weight ulvan was the major factor positively correlated with the gelling properties [67]. Ulvan from *Ulva fasciata* in different ionic strengths (Na^+ and Ca^{2+}) had significant effects on the stability of o/w emulsions [68]. The rheological properties and zeta-potential of the emulsions appeared to be dependent on the ulvan concentration. The emulsifying and stabilizing mechanism of the ulvan may ascribe to its surface-active protein moiety, also to the hydrophobicity of the polysaccharide itself [69].

3.3. Red seaweeds

Red seaweeds are usually grown or collected for food consumption and especially known for their health benefits and high nutritional value, such as nori or purple lave (*Porphyra* spp.), ogo, ogonori, or sea moss (*Gracilaria* spp.) The red seaweeds *Gracilaria* and *Gelidium* are used in the manufacture of the agar, *Kappaphycus* and *Betaphycus* are now the most important sources of carrageenan, and *Porphyra* could extract porphyran for food and biotechnological applications [70].

Carrageenan are high-molecular-weight sulfated D-galactans composed of repeating disaccharide units with alternating 3-linked β-D-galactopyranose and 4-linked α-galactopyranose or 3,6-anhydro-α-galactopyranose [7]. There are at least 15 different carrageenan structures, including but not limited to κ-, ι-, λ-, μ-, θ-, β-, and ν-carrageenan [20]. Porphyran has hypolipidemic and antiallergic pharmacological property applications, is a complex SWP, and is consist of about 30% of agarose repeating units (1,3-linked β-D-galactopyranose followed by 1,4-linked 3,6-anhydro-α-L-galactopyranose), with the remaining residues being 3-linked β-D-galactopyranose and 4-linked α-L-galactopyranose-6-sulfate. The composition includes 6-O-sulfated L-galactose, 6-O-methylated D-galactose, L-galactose, 3,6-anhydro-L-galactose, 6-O-methyl D-galactose, and ester sulfate. Some of the ester is present as 1-4-linked L-galactose 6-sulfate [71].

Kappa carrageenan showed that shear stress-shear rate and viscosity curves clearly indicated sharp increases in viscosity and consistency coefficient (k), and addition of KCl was more effective in increasing viscosity [72]. Kappa-carrageenan film incorporation of plant oils increased the film thickness and plasticizing effect significantly. However, the moisture content, solubility, and tensile strength of films decreased significantly when plant oils were added [73]. Six formulations carrageenan/porphyran films from *Pyropia columbina* were homogeneous and flexible. Their moisture content, water solubility, and water vapor permeability significantly increased by glycerol content increasing could obtain more stretchable but less resistant films. When Ca^{+2} addition could mask glycerol effect in water solubility and water vapor permeability, it could stabilize the three-dimensional structure of carrageenans/porphyrans by interactions between sulfate groups, promote water retention, and open film structure [74]. Porphyran, alkali-treated porphyrin, and sulfated porphyran from *Porphyra haitanensis* demonstrate that the anticoagulant activities mainly depend on the position of sulfate and the antioxidant activities mainly depend on degree of sulfate substitution [75]. The rheological behavior of porphyran exist pseudoplastic behavior, which agrees with the Herschel-Bulkley model. The effect of temperature on the viscosity of porphyrans was increasing concentrations from 3 to 7% [76].

4. Antioxidant activity of SWP

Reactive oxygen species (ROS) is produced in the forms of H_2O_2, superoxide radical ($O^{\cdot-2}$), hydroxyl radical (•OH) and nitric oxide (NO) in the organisms are produced by non-enzymatic and enzymatic reactions. Moreover, endogenous antioxidant enzymes and antioxidants in the body under synergy effect would been removed these ROS. But when the body is aging and experiences illness or fatigue, the body's free radicals may be destroyed, and excess oxygen free radicals cause a series of oxidative damage to the body. Excessive free radicals can cause irreversible damage to the body. It can cause damage to the body at molecular level, cell level, and level of tissue and organs by attacking the life molecules and various kinds of cells [77]. Oxidative stress and disease have resulted in identification of important oxidative stress biomarkers—the products of oxidation of biological molecules: DNA, lipids, and proteins. It would lead to many health disorders including cancer, diabetes, cardiovascular, inflammatory, and neurodegenerative diseases [78]. SWP from a number of seaweeds has appreciable antioxidant capability. Antioxidant activity of SWP has been determined by various in vitro methods, such as ABTS radical scavenging [79], 1,1-diphenyl-2-picrylhydrazyl (DPPH) radical scavenging [80], ferric reducing antioxidant power (FRAP) [81], lipid peroxide inhibition [82], NO scavenging [83], and superoxide and hydroxyl radical scavenging assays [84, 85].

4.1. SWP by EAE

Enzymes are more action specific and operate at lower temperature during hydrolysis; thus, EAE is more amicable for degradation of biological materials without damaging the bioactive compounds. The extract of *Ecklonia cava* (brown seaweed) obtained using Celluclast was compared with extract obtained by organic solvents. Methanol extract, which gave the highest radical scavenging activities among the organic solvents investigated, was 20% less than the extract obtained by Celluclast [86, 87].

EAE was performed using Alcalase, cellulase, flavourzyme, and Viscozyme L applied on *Sargassum muticum* (brown seaweed), *Codium tomentosum* (green seaweed), and *Osmundea pinnatifida* (red seaweed). SWP with higher extraction yields were observed for *Co. tomentosum* EAE (for cellulase and Viscozyme L), followed by *O. pinnatifida* (except Alcalase) and *S. muticum*. A higher effect on hydroxyl-radical scavenging activity (35–50%) was observed for all SWP extracts [88].

The extracts of *U. armoricana* (green seaweed) were determined to assess the efficiency of endo-protease treatments which significantly increased the extraction yields compared to the control. The organic matter, neutral sugar, and protein contents were increased in all extracts compared to an extraction with water, up to 2.0-, 2.7-, and 1.75-fold, respectively. Free radical scavenging capacity at medium inhibition concentrations (IC_{50}) of 1.8 and 12.5 mg/mL was shown for the extracts produced with endo-protease treatments and 6.0 mg/mL for the sample resulting from the extraction with the multiple mix of glycosyl hydrolases [89]. Enzyme-assisted extraction of *Un. pinnatifida* (green seaweed) was performed using five proteases (Alcalase, Flavourzyme, Neutrase, trypsin, and Protamex) and six carbohydrases (AMG, Dextrozyme, maltogenase, Promozyme, Viscozyme, and Celluclast). SWP of *Un. pinnatifida* exhibited strong radical scavenging activity on DPPH and hydroxyl radical, and activity increased with increasing concentration [90].

SWP were extracted from *Pterocladia capillacea* (red seaweed), and the produced fractions were hydrolyzed using different enzymes, such as Viscozyme L (a multienzyme complex containing arabanase, cellulase, β-glucanase, hemicellulase, and xylanase), β-glucanase from

Trichoderma longibrachiatum, and β-galactosidase from *Kluyveromyces lactis*. The Viscozyme hydrolysate exhibited DPPH scavenging activity of about 92% with more than 50% increase over its mother fraction [27]. Three SWP were obtained from the *Mastocarpus stellatus* (red seaweed) using Alcalase hydrolysis at three different treatments. The Alcalase enzymatic hydrolysis at 50°C provided films with reducing power and radical scavenging capacity, which did not change as a result of subsequent heat treatment at 90°C. Their viscoelastic properties of the film-forming solutions showed improved gel-like behavior when the κ/ι-hybrid carrageenan extraction at 90°C was promoted [91].

4.2. SWP by MAE

SWP (fucoidan) from *Ascophyllum nodosum* (brown seaweed) were extracted by MAE technology. Different conditions of temperature (90–150°C) and extraction time (5–30 min) were evaluated, and optimal fucoidan yield was 16.08%, obtained from 120°C for 15 min of extraction. All SWP from *A. nodosum* exhibited antioxidant activities as measured by DPPH scavenging and reducing power, among which SWP extracted at 90°C was the highest [62]. SWP were extracted from *Fucus vesiculosus* (brown seaweed) seaweed by using MAE. The SWP obtained by MAE that contained 53.8 mole% of fucose, 35.3 mole% of xylose, and 10.8 mole% of galactose presented comparable values of antioxidant activity by the DPPH, $ABTS^+$, and lipid oxidation inhibition methods [92].

Six representative (molecular weight 446.5, 247.0, 76.1, 19.0, 5.0, and 3.1 kDa) SWP from *Enteromorpha prolifera* (green seaweed) were extracted by MAE. All samples showed that great inhibitory effects on superoxide radical at a low concentration and high molecular weight exhibited higher inhibitory effects. Otherwise, samples with low molecular weight possessed stronger inhibitory effects on hydroxyl radical; IC_{50} of molecular weight 3.1 kDa was 0.39 mg/mL. The chelating effect of molecular weight 3.1 kDa was 77.3% at 5 mg/mL, which was twice more than initial polysaccharide [93].

Four SWP were extracted from *Durvillaea antarctica* (brown seaweed), *Sarcodia ceylonensis* (brown seaweed), *U. lactuca* L. (green seaweed), and *Gracilaria lemaneiformis* (red seaweed), respectively, by MAE. The average molecular weight of SWP of *D. antarctica*, *S. ceylonensis*, *U. lactuca* L., and *G. lemaneiformis* was 482, 466, 404, and 591 kDa, respectively. The in vitro antioxidant activity of all of the polysaccharides was evaluated using $ABTS^+$, hydroxyl radical, nitrite scavenging capacity, and reducing power. SWP of *U. lactuca* L. presented the highest ABTS radical scavenging activity; SWP of *U. lactuca* L., *S. ceylonensis*, and *D. antarctica* also showed a strong effect on the ABTS radical scavenging activity. SWP of *U. lactuca* L. and *S. ceylonensis* exhibited excellent hydroxyl-radical scavenging activities, about 83.33% ± 2.31% and 80.07% ± 2.17%, respectively, at 4 mg/mL. The reducing power of SWP of *D. Antarctica* was relatively more pronounced than that of the three other polysaccharides. However, the nitrite scavenging activities of the four seaweed polysaccharides were weaker than other antioxidant activities (ABTS), hydroxyl radical scavenging capacities, and reducing powers [94].

SWP from *Porphyra dentate* (red seaweed) is adjusting the pH value of the ethanol solutions used as extraction solvents and then applying continuous or intermittent MAE to extract *P. dentate* solutions. The SWP content was significantly affected by ethanol concentration, pH value of the ethanol solution, and intermittency at a 1% significance level. It also showed high antioxidant activities by the DPPH and FRAP. Gelling property of the extracted SWP was not affected [95].

4.3. SWP by PLE

PLE was utilized to extract SWP from *Saccharina japonica* (brown seaweed). Various conditions of temperature (80–200°C), pressure (5–100 bar), and solvents (water, 0.1% sodium hydroxide, 0.1% formic acid, 70% ethanol, 50% ethanol, and 25% ethanol) were assessed; the best crude SWP yield was 8.23%, obtained from 140°C and 50 bar (sodium hydroxide). All crude SWP showed antioxidant activities as measured by DPPH radical and $ABTS^+$ radical scavenging. Crude SWP demonstrates good emulsion-stabilizing capacities, especially with vegetable oils [96].

4.4. SWP by SFE

Gracilaria mammillaris (red seaweed) from the Colombian Caribbean coast were investigated as a source of extracts with antioxidant activity, by means of supercritical CO_2 extraction with ethanol as cosolvent. A central composite design was used to investigate the effects of pressure (10, 20, and 30 MPa), temperature (40, 50, and 60°C), and cosolvent concentration (2, 5, and 8%) on antioxidant activity. The antioxidant activity of samples was evaluated by determining their capacity for protecting an edible oil against oxidation, upon accelerated oxidation trials. The extracts obtained at 30 MPa, 60°C, and 8% cosolvent showed the highest antioxidant activity, inhibiting 42.1% in the formation of TBARS after 6 days of accelerated oxidation [97].

4.5. SWP by UAE

UAE was carried out in a water bath ultrasonicator for 60 min (sonicate for 10 min and pause for 2 min) at 50°C applied on *S. muticum* (brown seaweed), *Co. tomentosum* (green seaweed), and *O. pinnatifida* (red seaweed). A higher effect on hydroxyl-radical scavenging activity (35–50%) was observed for all SWP [88]. The SWP from *Gracilaria birdiae* (red seaweed) were obtained using five different extraction conditions. Their infrared and electrophoresis analysis showed that all conditions extracted the same SWP. The total capacity antioxidant of the SWP was also affected by extraction condition, since GB2s (NaOH/sonication/proteolysis extraction) and GB1 (water extraction) showed lower activity in comparison to the other conditions. The data revealed that NaOH/sonication/proteolysis was the best condition to extract antioxidant SWP from *G. birdiae* [98].

In general, the antioxidant properties of SWP are influenced by chemical characteristics like molecular weight, degree of branching, type of monosaccharides, ratio of monosaccharides, intermolecular associations of polysaccharides, glycosidic branching, and modification of polysaccharides. It has been observed that crude polysaccharide has better antioxidant activity than purified polysaccharide components. Antioxidant activity of SWP depends on their structural features, such as degree of sulfating, molecular weight, type of the major sugar, and glycosidic branching. The rationale is that low-molecular-weight SWP may incorporate into the cells more efficiently and donate protons more effectively than high-molecular-weight SWP. Evidence suggests that SWP are higher useful candidates when searching for effective nontoxic substances with potential antioxidant activity among various naturally occurring compounds. Therefore, SWP could be used as a rich source of natural antioxidants with potential application in the food industry as well as in cosmetic and pharmaceutical areas [99].

5. Conclusions

Seaweeds are used in many maritime countries as a source of human food, for industrial applications, and as a fertilizer. The major utilization of these plants as functional food is in Asia, particularly China, Japan, and Korea. They have the potential to be used as a source of long- and short-chain chemicals with medicinal and industrial uses. Seaweeds are a rich source of amino acids, bioactive peptides, carotenoids, dietary fiber, minerals, omega-3 fatty acids, SWP, and vitamins. Among them, SWP have the potential to significantly improve extraction efficiency. SWP include a complex group of macromolecules with a wide range of important biological activities, such as antioxidant, anticoagulant, anticancer, antiviral, anti-allergy, and anti-inflammation. Future research priorities in this area should be concentrated on overcoming the challenges of employing these novel technologies on chemical characteristics of SWP.

Author details

Shao-Chi Wu

Address all correspondence to: agarase@msn.com

Department of Food and Beverage Management, Tung-Fang Design University, Kaohsiung, Taiwan

References

[1] Roohinejad S, Koubaa M, Barba, FJ, Saljoughian S, Amid M, Greiner R. Application of seaweeds to develop new food products with enhanced shelf-life, quality and health-related beneficial properties. Food Research International. 2016. (In Press) DOI: 10.1016/j.foodres.2016.08.016

[2] Manivasagan P, Oh J. Marine polysaccharide-based nanomaterials as a novel source of nanobiotechnological applications. International Journal of Biological Macromolecules. 2016;**82**:315-327. DOI: 10.1016/j.ijbiomac.2015.10.081

[3] Fernando IP, Nah, JW, Jeon, YJ. Potential anti-inflammatory natural products from marine algae. Environmental Toxicology and Pharmacology. 2016;**48**:22-30. DOI: 10.1016/j.etap.2016.09.023

[4] McHugh DJ. A Guide to the Seaweed Industry; FAO Fisheries Technical Paper 441. Rome, Italy: Food and Agriculture Organization of the United Nations; 2003

[5] Fimbres-Olivarría D, López-Elías JA, Carvajal-Millán E, Márquez-Escalante JA, Martínez-Córdova LR, Miranda-Baeza A, Enríquez-Ocaña F, Valdéz-Holguín JE, Brown-Bojórquez F. *Navicula* sp. sulfated polysaccharide gels induced by Fe(III): Rheology and microstructure. International Journal of Molecular Sciences. 2016;**17**:1-10. DOI: 10.3390/ijms17081238

[6] Gurpilhares DB, Moreira TR, Bueno JL, Cinelli LP, Mazzola PG, Pessoa A, Sette LD. Algae's sulfated polysaccharides modifications: Potential use of microbial enzymes. Process Biochemistry. 2016;**51**:989-998. DOI: 10.1016/j.procbio.2016.04.020

[7] Cunha L, Grenha A. Sulfated seaweed polysaccharides as multifunctional materials in drug delivery applications. Marine Drugs. 2016;**14**:42. DOI: 10.3390/md14030042

[8] Di T, Chen GJ, Sun Y, Ou SY, Zeng XX, Hong Y. Antioxidant and immunostimulating activities *in vitro* of sulfated polysaccharides isolated from *Gracilaria rubra*. Journal of Functional Foods. 2017;**28**:64-75. DOI: 10.1016/j.jff.2016.11.005

[9] Wang JH, Zhang BW, Luo JP. Molecular weight, chain conformation and antioxidant activities of sulfated β-d-galactan derivatives from *Dendrobium nobile* Lindl. Current Topics in Nutraceutical Research. 2015;**13**:205-212

[10] Ameer K, Shahbaz HM, Kwon JH. Green extraction methods for polyphenols from plant matrices and their byproducts: A review. Comprehensive Reviews in Food Science and Food Safety. 2017;**16**:295-315. DOI: 10.1111/1541-4337.12253

[11] Kadam SU, Tiwari BK, O'Donnell CP. Application of novel extraction technologies for bioactives from marine algae. Journal of Agricultural and Food Chemistry. 2013;**61**:4667-4675. DOI: 10.1021/jf400819p

[12] Cheng X, Bi LG, Zhao ZD, Chen YX. Advances in enzyme assisted extraction of natural products. In: 3rd International Conference on Material, Mechanical and Manufacturing Engineering (IC3ME 2015); 27-28 June, 2015; Amstelkade, Amsterdam, Netherlands. Atlantis Press; 2015. pp. 371-375

[13] Wijesinghe WAJ, Jeon YJ. Enzyme-assistant extraction (EAE) of bioactive components: A useful approach for recovery of industrially important metabolites from seaweeds. Fitoterapia. 2012;**83**:6-12. DOI: 10.1016/j.fitote.2011.10.016

[14] Jiao GL, Yu GL, Zhang JZ, Ewart HS. Chemical structures and bioactivities of sulfated polysaccharides from marine algae. Marine Drugs. 2011;**9**:196-223. DOI: 10.3390/md9020196

[15] Michalak I, Chojnacka K. Algal extracts: Technology and advances. Engineering in Life Sciences. 2014;**14**:581-591. DOI: 10.1002/elsc.201400139

[16] Lee SH, Athukorala Y, Lee JS, Jeon YJ. Simple separation of anticoagulant sulfated galactan from marine red algae. Journal of Applied Phycology. 2008;**20**:1053-1059. DOI: 10.1007/s10811-007-9306-0

[17] Sila A, Bougatef A. Antioxidant peptides from marine by-products: Isolation, identification and application in food systems. A review. Journal of Functional Foods. 2016;**21**:10-26. DOI: 10.1016/j.jff.2015.11.007

[18] Cao RA, Lee YJ, You SG. Water soluble sulfated-fucans with immune-enhancing properties from *Ecklonia cava*. International Journal of Biological Macromolecules. 2014;**67**:303-311. DOI: 10.1016/j.ijbiomac.2014.03.019

[19] Hahn T, Lang S, Ulber R, Muffler K. Novel procedures for the extraction of fucoidan from brown algae. Process Biochemistry. 2012;**47**:1691-1698. DOI: 10.1016/j.procbio.2012.06.016

[20] Hardouin K, Burlot AS, Umami A, Tanniou A, Stiger-Pouvreau V, Widowati I, Bedoux G, Bourgougnon N. Biochemical and antiviral activities of enzymatic hydrolysates from different invasive French seaweeds. Journal of Applied Phycology. 2014;**26**:1029-1042. DOI: 10.1007/s10811-013-0201-6

[21] Fleita D, El-Sayed M, Rifaat D. Evaluation of the antioxidant activity of enzymatically-hydrolyzed sulfated polysaccharides extracted from red algae; *Pterocladia capillacea*. LWT - Food Science and Technology. 2015;**63**:1236-1244. DOI: 10.1016/j.lwt.2015.04.024

[22] Chen Y, Yao FK, Ming K, Wang DY, Hu YL. Polysaccharides from traditional Chinese medicines: Extraction, purification, modification, and biological activity. Molecules. 2016;**21**:1-23. DOI: 10.3390/molecules21121705

[23] Hardouin K, Bedoux G, Burlot AS, Navall-Collen P, Bourgougnon N. Enzymatic recovery of metabolites from seaweeds: Potential applications. In: Bourgougnon N, editor. Advances in Botanical Research. Vol. 71 Sea Plants. UK: Elsevier; 2014

[24] Tatke P, Jaiswal Y. An overview of microwave assisted extraction and its applications in herbal drug research. Research Journal of Medicinal Plant. 2011;**5**:21-31. DOI: 10.3923/rjmp.2011.21.31

[25] Delazar A, Nahar L, Hamedeyazdan S, Sarker SD. Microwave-assisted extraction in natural products isolation. Methods in Molecular Biology. 2012;**864**:89-115. DOI: 10.1007/978-1-61779-624-1_5

[26] Chan CH, Yeoh HK, Yusoff R, Ngoh GC. A first-principles model for plant cell rupture in microwave-assisted extraction of bioactive compounds. Journal of Food Engineering. 2016;**188**:98-107. DOI: 10.1016/j.jfoodeng.2016.05.017

[27] Chin CL. Study of Extraction Processes and their Impact on Bioactivity of Botanicals. A thesis submitted for the degree of doctor of philosophy department of pharmacy national university of Singapore; 2009

[28] Halfadji A, Touabet A. Badjah-Hadj-Ahmed AY. Comparison of soxhlet extraction, microwave-assisted extraction and ultrasonic extraction for the determination of PCBs congeners in spiked soils by transformer oil (Askarel). International Journal of Advances in Engineering and Technology. 2013;**5**:63-75

[29] Sookjitsumran W, Devahastin S, Mujumdar AS, Chiewchan N. Comparative evaluation of microwave-assisted extraction and preheated solvent extraction of bioactive compounds from a plant material: A case study with cabbages. International Journal of Food Science and Technology. 2016;**51**:2440-2449. DOI: 10.1111/ijfs.13225

[30] Patil DM, Akamanchi KG. Microwave assisted process intensification and kinetic modelling: Extraction of camptothecin from *Nothapodytes nimmoniana* plant. Industrial Crops and Products. 2017;**98**:60-67. DOI: 10.1016/j.indcrop.2017.01.023

[31] Sun Y, Xue HK, Liu CH, Liu C, Su XL, Zheng XZ. Comparison of microwave assisted extraction with hot reflux extraction in acquirement and degradation of anthocyanin from powdered blueberry. International Journal of Agricultural and Biological Engineering. 2016;**9**:186-199. DOI: 10.3965/j.ijabe.20160906.2724

[32] Czarnecki S, During RA. Closed-vessel miniaturised microwave-assisted EDTA extraction to determine trace metals in plant materials. International Journal of Environmental Analytical Chemistry. 2014;**94**:801-811

[33] Quoc LPT, van Muoi N. Microwave-assisted extraction of phenolic compounds from *Polygonum multiflorum* Thunb. roots. Acta Scientiarum Polonorum. Technologia Alimentaria. 2016;**15**:181-189. DOI: 10.17306/J.AFS.2016.2.18

[34] Zhang HF, Yang XH, Wang Y. Microwave assisted extraction of secondary metabolites from plants: Current status and future directions. Trends in Food Science and Technology. 2011;**22**:672-688. DOI: 10.1016/j.tifs.2011.07.003

[35] Yanik DK. Alternative to traditional olive pomace oil extraction systems: Microwave-assisted solvent extraction of oil from wet olive pomace. LWT - Food Science and Technology. 2017;**77**:45-51. DOI: 10.1016/j.lwt.2016.11.020

[36] Welna M, Borkowska-Burnecka J, Popko M. Ultrasound- and microwave-assisted extractions followed by hydride generation inductively coupled plasma optical emission spectrometry for lead determination in geological samples. Talanta. 2015;**144**:953-959. DOI: 10.1016/j.talanta.2015.07.058

[37] Benthin B, Danz H, Hamburger M. Pressurized liquid extraction of medicinal plants. Journal of Chromatography A. 1999;**837**:211-219. DOI: 10.1016/S0021-9673(99)00071-0

[38] Osorio-Tobón JF, Meireles MAA. Recent applications of pressurized fluid extraction: Curcuminoids extraction with pressurized liquids. Food and Public Health. 2013;**3**:289-303. DOI: 10.5923/j.fph.20130306.05

[39] Wu HZ, Wang JM, Yang H, Li GQ, Zeng YH, Xia W, Li ZG, Qian MR. Development and application of an in-cell cleanup pressurized liquid extraction with ultra-high-performance liquid chromatography-tandem mass spectrometry to detect prohibited antiviral agents sensitively in livestock and poultry feces. Journal of Chromatography A. 2017;**1488**:10-16. DOI: 10.1016/j.chroma.2017.01.070

[40] Chen HP, Pan ML, Liu X, Lu CY. Evaluation of transfer rates of multiple pesticides from green tea into infusion using water as pressurized liquid extraction solvent and

ultra-performance liquid chromatography tandem mass spectrometry. Food Chemistry. 2017;**216**:1-9. DOI: 10.1016/j.foodchem.2016.07.175

[41] Klees M, Bogatzki C, Hiester E. Selective pressurized liquid extraction for the analysis of polychlorinated biphenyls, polychlorinated dibenzo-p-dioxins and dibenzofurans in soil. Journal of Chromatography A. 2016;**1468**:10-16. DOI: 10.1016/j.chroma.2016.09.029

[42] Jiménez-Salcedo M, Tena MT. Determination of cinnamaldehyde, carvacrol and thymol in feedstuff additives by pressurized liquid extraction followed by gas chromatography-mass spectrometry. Journal of Chromatography A. 2017;**1487**:14-21. DOI: 10.1016/j.chroma.2017.01.042

[43] Rai A, Mohanty B, Bhargava R. Fitting of broken and intact cell model to supercritical fluid extraction (SFE) of sunflower oil. Innovative Food Science and Emerging Technologies. 2016;**38**:32-40. DOI: 10.1016/j.ifset.2016.08.019

[44] Salea R, Veriansyah B, Tjandrawinata RR. Optimization and scale-up process for supercritical fluids extraction of ginger oil from *Zingiber officinale* var. Amarum. The Journal of Supercritical Fluids. 2017;**120**:285-294. DOI: 10.1016/j.supflu.2016.05.035

[45] Chen MH, Huang TC. Volatile and nonvolatile constituents and antioxidant capacity of oleoresins in three Taiwan citrus varieties as determined by supercritical fluid extraction. Molecules. 2016;**21**:1-12. DOI: 10.3390/molecules21121735

[46] Osorio-Tobón JF, Carvalho PIN, Rostagno MA, Meireles MAA. Process integration for turmeric products extraction using supercritical fluids and pressurized liquids: Economic evaluation. Food and Bioproducts Processing: Transactions of the Institution of Chemical Engineers Part C. 2016;**98**:227-235. DOI: 10.1016/j.fbp.2016.02.001

[47] Antunes-Ricardo M, Gutiérrez-Uribe JA, Guajardo-Flores D. Extraction of isorhamnetin conjugates from *Opuntia ficus-indica* (L.) Mill using supercritical fluids. The Journal of Supercritical Fluids. 2017;**119**:58-63. DOI: 10.1016/j.supflu.2016.09.003

[48] Picó Y. Ultrasound-assisted extraction for food and environmental samples. Trends in Analytical Chemistry. 2013;**43**:84-99. DOI: 10.1016/j.trac.2012.12.005

[49] Yilmaz T, Kumcuoglu S, Tavman S. Ultrasound-assisted extraction of lycopene and β-carotene from tomato-processing wastes. Italian Journal of Food Science. 2017;**29**:186-194

[50] Zhang LJ, Wang MS. Polyethylene glycol-based ultrasound-assisted extraction and ultrafiltration separation of polysaccharides from *Tremella fuciformis* (snow fungus). Food and Bioproducts Processing: Transactions of the Institution of Chemical Engineers Part C. 2016;**100**:464-468. DOI: 10.1016/j.fbp.2016.09.007

[51] Abdelhedi O, Nasri R, Souissi N, Nasri M, Jridi M. Sulfated polysaccharides from common smooth hound: Extraction and assessment of anti-ACE, antioxidant and antibacterial activities. Carbohydrate Polymers. 2016;**152**:605-614. DOI: 10.1016/j.carbpol.2016.07.048

[52] Netanel Liberman G, Ochbaum G, (Malis) Arad S, Bitton R. The sulfated polysaccharide from a marine red microalga as a platform for the incorporation of zinc ions. Carbohydrate Polymers. 2016;**152**:658-664. DOI: 10.1016/j.carbpol.2016.07.025

[53] Tang L, Chen YC, Jiang ZB, Zhong SP, Chen WZ, Zheng FC, Shi GG. Purification, partial characterization and bioactivity of sulfated polysaccharides from *Grateloupia livida*. International Journal of Biological Macromolecules. 2017;**94**:642-652. DOI: 10.1016/j.ijbiomac.2016.10.067

[54] Ngo DH, Kim SK. Sulfated polysaccharides as bioactive agents from marine algae. International Journal of Biological Macromolecules. 2013;**62**:70-75. DOI: 10.1016/j.ijbiomac.2013.08.036

[55] Lu KY, Li R, Hsu CH, Lin CW, Chou SC, Tsai ML, Mi FL. Development of a new type of multifunctional fucoidan-based nanoparticles for anticancer drug delivery. Carbohydrate Polymers. 2017;**165**:410-420. DOI: 10.1016/j.carbpol.2017.02.065

[56] Fletcher HR, Biller P, Ross AB, Adams JMM. The seasonal variation of fucoidan within three species of brown macroalgae. Algal Research. 2017;**22**:79-86. DOI: 10.1016/j.algal.2016.10.015

[57] Shchipunov YA, Mukhaneva OG, Zvyagintseva TN, Popivnich IB, Shevchenko NM. Rheological properties of aqueous fucoidan solutions. Vysokomolekulyarnye Soedineniya Seriya A and Seriya B. 2000;**42**:93-101

[58] Tako M. Rheological characteristics of fucoidan isolated from commercially cultured *Cladosiphon okamuranus*. Botanica Marina. 2003;**46**:461-465

[59] Kim DY, Shin WS. Functional improvements in bovine serum albumin–fucoidan conjugate through the Maillard reaction. Food Chemistry. 2016;**190**:974-981. DOI: 10.1016/j.foodchem.2015.06.046

[60] Hifney AF, Fawzy MA, Abdel-Gawad KM, Gomaa M. Industrial optimization of fucoidan extraction from *Sargassum* sp. and its potential antioxidant and emulsifying activities. Food Hydrocolloids. 2016;**54**:77-88. DOI: 10.1016/j.foodhyd.2015.09.022

[61] Wang LC, Wang XY, Wu H, Liu R. Overview on biological activities and molecular characteristics of sulfated polysaccharides from marine green algae in recent years. Marine Drugs. 2014;**12**:4984-5020. DOI: 10.3390/md12094984

[62] Darah I, Tong WY, Nor-Afifah S, Nurul-Aili Z, Lim SH. Antimicrobial effects of *Caulerpa sertularioides* extract on foodborne diarrhea-caused bacteria. European Review for Medical and Pharmacological Sciences. 2014;**18**:171-178

[63] Li N, Mao WJ, Yan MX, Liu X, Xia Z, Xia Z, Wang SY, Xiao B, Chen CL, Zhang LF, Cao SJ. Structural characterization and anticoagulant activity of a sulfated polysaccharide from the green alga *Codium divaricatum*. Carbohydrate Polymers. 2015;**121**:175-182. DOI: 10.1016/j.carbpol.2014.12.036

[64] Harada N, Maeda M. Chemical structure of antithrombin-active rhamnan sulfate from *Monostrom nitidum*. Bioscience, Biotechnology, and Biochemistry. 1998;**62**:1647-1652

[65] Thanh TTT, Quach TMT, Nguyen TN, Vu Luong D, Bui ML, Tran TTV. Structure and cytotoxic activity of ulvan extracted from green seaweed *Ulva lactuca*. International Journal of Biological Macromolecules. 2016;**93**:695-702. DOI: 10.1016/j.ijbiomac.2016.09.04

[66] Tsubaki S, Hiraoka M, Hadano S, Nishimura H, Kashimura K. Functional group dependent dielectric properties of sulfated hydrocolloids extracted from green macroalgal biomass. Carbohydrate Polymer. 2014;**107**:192-197. DOI: 10.1016/j.carbpol.2014.03.002

[67] Robic A, Sassi JF, Dion P, Lerat Y, Lahaye M. Seasonal variability of physicochemical and rheological properties of ulvan in two *Ulva* species (Chlorophyta) from the Brittany coast. Journal of Phycology. 2009;**45**:962-973

[68] Shao P, Ma HL, Zhu JY, Qiu Q. Impact of ionic strength on physicochemical stability of o/w emulsions stabilized by *Ulva fasciata* polysaccharide. Food Hydrocolloids. 2017;**69**:202-209. DOI: 10.1016/j.foodhyd.2017.01.039

[69] Shao P, Shao JM, Jiang YK, Sun PL. Influences of *Ulva fasciata* polysaccharide on the rheology and stabilization of cinnamaldehyde emulsions. Carbohydrate Polymers. 2016;**135**:27-34. DOI: 10.1016/j.carbpol.2015.08.075

[70] Nishiguchi T, Cho K, Isaka S, Ueno M, Jin J-O, Yamaguchi K, Kim DK, Oda T. Protective effect of porphyran isolated from discolored nori (*Porphyra yezoensis*) on lipopolysaccharide-induced endotoxin shock in mice. International Journal of Biological Macromolecules. 2016;**93**:1273-1278. DOI: 10.1016/j.ijbiomac.2016.09.091

[71] Bhatia S, Rathee P, Sharma K, Chaugule BB, Kar N, Bera T. Immuno-modulation effect of sulphated polysaccharide (porphyran) from *Porphyra vietnamensis*. International Journal of Biological Macromolecules. 2013;**57**:50-56. DOI: 10.1016/j.ijbiomac.2013.03.01

[72] Azizi R, Farahnaky A. Ultrasound assisted-viscosifying of kappa carrageenan without heating. Food Hydrocolloids. 2016;**61**:85-91. DOI: 10.1016/j.foodhyd.2016.05.006

[73] Nur Fatin Nazurah R, Nur Hanani ZA. Physicochemical characterization of kappa-carrageenan (*Euchema cottoni*) based films incorporated with various plant oils. Carbohydrate Polymers. 2017;**157**:1479-1487. DOI: 10.1016/j.carbpol.2016.11.026

[74] Cian RE, Salgado PR, Drago SR, Mauri AN. Effect of glycerol and Ca^{+2} addition on physicochemical properties of edible carrageenan/porphyran-based films obtained from the red alga, *Pyropia columbina*. Journal of Applied Phycology. 2015;**27**:1699-1708

[75] Zhang ZS, Zhang QB, Wang J, Song H, Zhang H, Niu XZ. Regioselective syntheses of sulfated porphyrans from *Porphyra haitanensis* and their antioxidant and anticoagulant activities *in vitro*. Carbohydrate Polymers. 2010;**79**:1124-1129. DOI: 10.1016/j.carbpol.2009.10.055

[76] In SK, Koo JG. Chemical composition and rheological properties of enzymatic hydrolysate of porphyran isolated from *Pyropia yezoensis*. Korean Journal of Fisheries and Aquatic Sciences. 2015;**48**:58-63. DOI: 10.5657/KFAS.2015.0058

[77] Shu GW, Zhang BW, Zhang Q, Wan HC, Li H. Effect of temperature, pH, enzyme to substrate ratio, substrate concentration and time on the antioxidative activity of hydrolysates from goat milk casein by alcalase. Acta Universitatis Cinbinesis, Series E: Food Technology. 2016;**20**:29-38. DOI: 10.1515/aucft-2016-0013

[78] Dhingra N, Kar A, Sharma R, Bhasin S. *In-vitro* antioxidative potential of different fractions from *Prunus dulcis* seeds: Vis a vis antiproliferative and antibacterial activities of active compounds. South African Journal of Botany. 2017;**108**:184-192. DOI: 10.1016/j.sajb.2016.10.013

[79] Gómez-Ordóñez E, Jiménez-Escrig A, Rupérez P. Bioactivity of sulfated polysaccharides from the edible red seaweed *Mastocarpus stellatus*. Bioactive Carbohydrates and Dietary Fibre. 2014;**3**:29-40. DOI: 10.1016/j.bcdf.2014.01.002

[80] Seedevi P, Moovendhan M, Viramani S, Shanmugam A. Bioactive potential and structural characterization of sulfated polysaccharide from seaweed (*Gracilaria corticata*). Carbohydrate Polymers. 2017;**155**:516-524. DOI: 10.1016/j.carbpol.2016.09.0

[81] Xie JH, Wang ZJ, Shen MY, Nie SP, Gong B, Li HS, Zhao Q, Li WJ, Xie MY. Sulfated modification, characterization and antioxidant activities of polysaccharide from *Cyclocarya paliurus*. Food Hydrocolloids. 2016;**53**:7-15. DOI: 10.1016/j.foodhyd.2015.02.018

[82] Ben Gara A, Ben Abdallah Kolsi R, Chaaben R, Hammami N, Kammoun M, Paolo Patti F, El Feki A, Fki L, Belghith H, Belghith K. Inhibition of key digestive enzymes related to hyperlipidemia and protection of liver-kidney functions by *Cystoseira crinita* sulphated polysaccharide in high-fat diet-fed rats. Biomedicine and Pharmacotherapy. 2017;**85**:517-526. DOI: 10.1016/j.biopha.2016.11.059

[83] Badrinathan S, Shiju TM, Christa ASS, Arya R, Pragasam V. Purification and structural characterization of sulfated polysaccharide from *Sargassum myriocystum* and its efficacy in scavenging free radicals. Indian Journal of Pharmaceutical Sciences. 2013;**74**:549-555

[84] Abu R, Jiang Z, Ueno M, Okimura T, Yamaguchi K, Oda T. *In vitro* antioxidant activities of sulfated polysaccharide ascophyllan isolated from *Ascophyllum nodosum*. International Journal of Biological Macromolecules. 2013;**59**:305-312. DOI: 10.1016/j.ijbiomac.2013.04.035

[85] Deng C, Xu JJ, Fu HT, Chen JH, Xu X. Characterization, antioxidant and cytotoxic activity of sulfated derivatives of a water-insoluble polysaccharides from *Dictyophora indusiata*. Molecular Medicine Reports. 2015;**11**:2991-2998. DOI: 10.3892/mmr.2014.3060

[86] Heo SJ, Lee KW, Song CB, Jeon YJ. Antioxidant activity of enzymatic extracts from brown seaweeds. Bioresource Technology. 2005;**96**:1613-1623. DOI: 10.1016/j.biortech.2004.07.013

[87] Heo SJ, Jeon YJ, Lee J, Kim HT, Lee KW. Antioxidant effect of enzymatic hydrolyzate from a Kelp, *Ecklonia cava*. Algae. 2003;**18**:341-347. DOI: 10.4490/ALGAE.2003.18.4.341

[88] Rodrigues D, Sousa S, Silva A, Amorim M, Pereira L, Rocha-Santos TAP, Gomes AMP, Duarte AC, Freitas AC. Impact of enzyme- and ultrasound-assisted extraction methods on biological properties of red, brown, and green seaweeds from the central west coast of Portugal. Journal of Agricultural and Food Chemistry. 2015;**63**:3177-3318

[89] Hardouin K, Bedoux G, Burlot AS, Donnay-Moreno C, Bergé JP, Nyvall-Collén P, Bourgougnon N. Enzyme-assisted extraction (EAE) for the production of antiviral and

antioxidant extracts from the green seaweed *Ulva armoricana* (Ulvales, Ulvophyceae). Algal Research. 2016;**16**:233-239. DOI: 10.1016/j.algal.2016.03.013

[90] Je JY, Park PJ, Kim EK, Park JS, Yoon HD, Kim KR, Ahn CB. Antioxidant activity of enzymatic extracts from the brown seaweed *Undaria pinnatifida* by electron spin resonance spectroscopy. LWT - Food Science and Technology. 2009;**42**:874-878. DOI: 10.1016/j.lwt.2008.10.012

[91] Alemán A, Blanco-Pascual N, Montero MP, Gómez-Guillén MC. Simple and efficient hydrolysis procedure for full utilization of the seaweed *Mastocarpus stellatus* to produce antioxidant films. Food Hydrocolloids. 2016;**56**:277-284. DOI: 10.1016/j.foodhyd.2015.12.024

[92] Rodriguez-Jasso R, Mussatto S, Pastrana L, Aguilar C, Teixeira J. Chemical composition and antioxidant activity of sulphated polysaccharides extracted from *Fucus vesiculosus* using different hydrothermal processes. Chemical Papers. 2014;**68**:203-209

[93] Li B, Liu S, Xing RG, Li KC, Li RF, Qin YK, Wang XQ, Wei ZH, Li PC. Degradation of sulfated polysaccharides from *Enteromorpha prolifera* and their antioxidant activities. Carbohydrate Polymers. 2013;**92**:1991-1996. DOI: 10.1016/j.carbpol.2012.11.088

[94] He JZ, Xu YY, Chen HB, Sun PL. Extraction, structural characterization, and potential antioxidant activity of the polysaccharides from four seaweeds. International Journal of Molecular Sciences. 2016;**17**:1-17. DOI: 10.3390/ijms17121988

[95] Wu SC, Lin YP, King VAE. Optimization of intermittent microwave-assisted extraction of sulfated porphyran from *Porphyra dentate*. Transactions of the ASABE. 2014;**57**:103-110

[96] Saravana PS, Cho YJ, Park YB, Woo HC, Chun BS. Structural, antioxidant, and emulsifying activities of fucoidan from *Saccharina japonica* using pressurized liquid extraction. Carbohydrate Polymers. 2016;**153**:518-525. DOI: 10.1016/j.carbpol.2016.08.014

[97] Ospina M, Castro-Vargas HI, Parada-Alfonso F. Antioxidant capacity of Colombian seaweeds: 1. Extracts obtained from *Gracilaria mammillaris* by means of supercritical fluid extraction. The Journal of Supercritical Fluids. 2017. (In Press) DOI: 10.1016/j.supflu.2017.02.023

[98] Fidelis GP, Gomes Camara RB, Queiroz MF, Santos Pereira Costa MS, Santos PC, Oliveira Rocha HA, Costa LS. Proteolysis, NaOH and ultrasound-enhanced extraction of anticoagulant and antioxidant sulfated polysaccharides from the edible seaweed, *Gracilaria birdiae*. Molecules. 2014;**19**:18511-18526

[99] Li SQ, Dai SH, Shah NP. Sulfonation and antioxidative evaluation of polysaccharides from *Pleurotus mushroom* and *Streptococcus thermophilus* bacteria: A review. Comprehensive Reviews in Food Science and Food Safety. 2017;**16**:282-294. DOI: 10.1111/1541-4337.12252

Chapter 7

Solubility of Chitin: Solvents, Solution Behaviors and Their Related Mechanisms

Jagadish C. Roy, Fabien Salaün, Stéphane Giraud,
Ada Ferri, Guoqiang Chen and Jinping Guan

Additional information is available at the end of the chapter

http://dx.doi.org/10.5772/intechopen.71385

Abstract

Chitin is a natural polysaccharides having a unique molecular arrangement of 2-(acetylamino)-2-deoxy-D-glucose, it possesses multifunctional properties and is suitable for various applications mainly in pharmaceutical, biomedical food, textiles and packaging fields. Therefore, being considered as a superior material for a sustainable future of industrial development, chitin perfectly meets up the demands with diversified functionalities in applications, excellent biocompatibility and biodegradability. Non-toxicity to human and environment (air, water and soil) is a great opportunity for this revolutionary, innovative and sustainable material. Moreover, antibacterial potency and low immunogenicity of chitin have broadened the aspects of research and development on structure-function relationship toward biological tissues and activities. Despite abundance, low cost and availability, many experimental data from potential studies, reproducibility problems of chitin solubility measurement still limit the development of products and access to the market in large volume. Batch-to-batch variability, non-precise characterization and randomly distributed acetyl groups of chitin structure eventually results in a bad reproducibility of chitin solubility. Therefore, the chapter aims to organize the information of chitin structure at molecular level and correlate solubility with chitin structure. Moreover, the dissolution mechanism and solution behaviors in different solvents will be discussed in this chapter.

Keywords: polysaccharide, chitin, chitosan, solubility, dissolution, hydrolysis

1. Introduction

Chitin is a polysaccharide consisting of glycosidic bonds in linear or branched fashion between two adjacent monosaccharides, 2-(acetylamino)-2-deoxy-D-glucose. In general, monosaccharides undergo a polycondensation reaction to link more than 20 units of oligosaccharides by glycosidic linkages. Most polysaccharides show the degree of average polymerization (number average DP) around 200–3000 while longer polysaccharide (like cellulose) exhibits DP around 7000–15,000. The presence of acetyl, amino and hydroxyl groups in the polysaccharide chain, due to the generation of hydrogen bonds (inter and intramolecular) makes the chitin highly aggregated. Therefore, it is insoluble in all regular solvents such as water, organic solvents, mild acidic or basic solution, etc. Chitin insolubility affects the scaling up of the processes for the production of chitin-based products. The first study on solubility determination of chitin was done by Austin who tested chitin in different solvents [1]. It was a well-organized evaluation of chitin solubility in different types of solvents such as dichloroacetic (DCA) and trichoroacetic (TCA) acids in presence or absence of alcohol, etc. Later on, many studies were conducted with the same intention by many other scientists and chitin solubility was verified in many solvents such as dimethylacetamide (DMA)/LiCl mixture [2], $CaBr_2{\cdot}H_2O$ saturated methanol [3], hexafluoroisopropyl alcohol and hexafluoraceton [4], lithium thiocyanate [5], phosphoric acid [6] and N-methyl-2-pyrrolidone [7], etc. Although dissolution of chitin is possible by these solvents many of them are toxic, scarcely degradable, corrosive, or mutagenic. Therefore, the choice of an appropriate solvent for chitin and chitosan solubilization is important and primary issue for lab scale research and scaling up for industrial practices. The acetyl groups in chitin can be removed by deacetylation to convert insoluble chitin into a more soluble compound, namely chitosan (this name is given to chitin with at least 50% degree of deacetylation, DD). Therefore, this chapter will deal with the deacetylation process and the changes in molecular orientations and bonds after the deacetylation reaction. Moreover, the chapter will revise technical details regarding different aspects of solution behavior of chitin and chitosan. The parameters influencing solubility and their action mechanisms, theoretical discussions and recent relevant research findings on chitin and chitosan will be found in this chapter. Finally, modification of polysaccharides and their enhanced solubility will be discussed.

2. Molecular structure of chitin and chitosan

Polysaccharides mimic protein and amino acids structure consisting of special conformation of secondary, tertiary and quaternary architectural structures. Chitin is arranged as crystalline microfibrils clustered with six-stranded helixes of a protein structure. Polymerization of the monosaccharides, β-(1-4)-2-acetamido-2-deoxy-β-ᴅ-glucosamine units exhibit three different polymorphism (α, β and γ sheets) whereas the N-acetyl glycosyl moiety is a common crystallographic unit in all [2]. Electron diffraction studies reveal a highly crystalline nature of chitin in the α and β conformation. The α-conformation is one of the most abundant allomorphs in which the unit polymer chains are arranged in an antiparallel fashion whereas the adjacent chains are always in the opposite direction [3]. In β, less frequent in nature sheet all chains are in the same direction and parallel, γ conformation is a variant of the α arrangement in which two parallel, adjacent and unidirectional chains are arranged with one opposite directional chain (**Figure 1**). Both α and β conformations maintain a strong network dominated by intra-chain hydrogen bonds between the groups of C=O···NH and C=O···OH within a distance of 0.47 nm. In α-chitin conformation, additional inter-chain hydrogen bonds bind the hydroxymethyl groups while this type of interaction is not observed in the β conformation. Thus,

http://dx.doi.org/10.5772/intechopen.71385

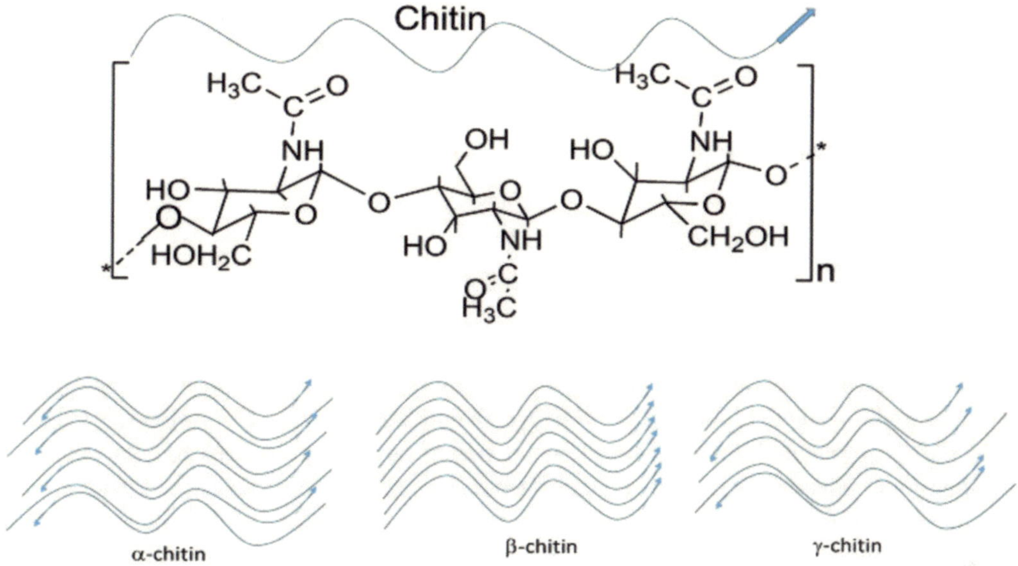

Figure 1. Structure of different chitin conformations (α, β and γ-chitin domain conformation).

β-chitin conformation is more prone to intra-crystalline swelling than α-chitin conformation. Different structural arrangement of α-chitin and β-chitin provide the reason why water, alcohol and amines are able to get access to the β conformation by swelling with and without disrupting the chain arrangement and crystalline structure [4]. For example, the swelling of β sheets in concentrated acidic solution of HNO_3 or HCl solution exhibits a permanent transformation into α-chitin conformation. In this case, a partial dissolution occurs driven by hydrolysis induced by the acid treatment. The recrystallization process of the smaller hydrolyzed chain starts on the un-hydrolyzed chitin sheets, which are called "epitaxy". Therefore, no single crystal growth is observed during the recrystallization and new α-chitin conformation crystals are observed. The transformation β→α indicates that the α-chitin sheet is thermodynamically stable and stability is achieved only via recrystallization [3]. The α-chitin sheets are not swelled by water and alcohol while aliphatic diamine or highly basic solutions can diffuse into the crystalline structure to promote the formation of chain complexes. The parameter value of inter-chain expansion is same as β-sheet around 0.7 nm. Therefore, a simple processing of β-chitin with 20% NaOH results into the conversion of α to β-conformations.

Chitin polysaccharides contain functional amino groups in its backbone to provide positively charged polysaccharide upon solubilization. The amount of reactive amino groups can be increased by increasing deacetylation which is quantified by the degree of deacetylation (%DD). Chitin is the only positively charged polysaccharide among all other naturally occurred biopolymers which allows a wide range of biological applications. There are two main groups in the chitin structure influences the functionality of chitin: (i) amino groups and (ii) hydroxyl groups (**Figure 1**). The amino sites might react with aldehyde and ketone groups for the Schiff Bases formation and influence solubility. In addition, two hydroxyl groups in chitin structure provide excellent pathways for modification and functionalization in view of an increase of solubility. Those hydroxyl groups involve in the O-acetylation, O-alkylation, H-bonds formation, etc. [5]. Also, the amino groups are responsible for the short-range primary and secondary electrostatic interaction while the second one involves the formation of hydrogen bonds. Moreover, the available unhydrolysed acetyl groups in chitin molecules form hydrophobic bonds in the solution and get aggregated [6].

3. Factors affecting the solubility of chitin

3.1. Effect of N-acetyl-D-glucosamine units

Chitins may have different acetylation depending on the sources such as fungi, insects, crustaceans or molluscs. Due to its crystalline structure with strong hydrogen bonds and cohesive forces, highly aggregated three-dimensional network is developed which leads to insolubility in conventional solvents. Pure chitin contains around 90% N-acetyl groups in its backbone and some deacetylation reactions take place due to the extraction process of chitin from the natural sources. There are two monomer units present in the chitin structure in different fraction: (i) 2-acetamino-2-deoxy-D-glucopyronase (N-acetyl-D-glucosamine) and (ii) 2-amino-2-deoxy-D-glucopyronase (N-amino-D-glucosamine). The first one, 2-acetamino-2-deoxy-D-glucopyronase, displays insolubility due to the strong hydrogen bonds between the acetyl groups of the same or adjacent chitin chains. Hydrogen bonds network builds a three-dimensional crystalline matrix by sequencing the following bonds —NH···O═C and —OH···O═C. The other unit, N-amino-D-glucosamine shows a distinct property such as hydrophilic nature and positively charged in acidic solution. The domination of the hydrophilic character with a high amount of N-amino-D-glucosamine unit in the chitin backbone can be determined by degrees of deacetylation (DD). The DD is determined from the ratio of N-amino-D-glucosamine to N-acetyl-D-glucosamine while the degree of acetylation (DA) represents the deduction from 100 (i.e. 100—DD). When DD is between 60 and 90% a new chemical entity "Chitosan" is baptized which is soluble in organic acids such as acetic acid. Alternatively, the structure with a DD value less than 60% is regarded as chitin and insoluble in acidic solutions. Chitin is treated with alkali solution (NaOH) for deacetylation to occur. The type of deacetylation process provides a different distribution of acetyl groups in the chitin backbone: micro-block and random. The micro-block domain chitosan is easily susceptible to aggregation along the extended series of acetyl units which leads to a complete insolubility in the majority of the common solvents. In the case of high DA > 60% or DD < 40%, chitosan is very prone to association and aggregation while chitin with low DA = 1.5% or DD = 98.5 did not show any aggregation [7]. Simina and coworkers reported the aggregation behavior for a wide range of DD at pH 4.5 and determined the hydrodynamic radius at DD – 98.5, 63 and 31%. For very high DD (>75%), protonated charge condensation occurs in the chitosan solution due to large charge density which leads to electrostatic repulsion and high solubility. As a result, single size distribution of chitin molecules was observed with an average diameter of 40 nm at DD 98.5% [7]. When the DD was moderated (75–50%), an additional size distribution of 300 nm (an average diameter) with the previous size distribution (40 nm) was displayed which indicated the aggregated form of chitin. The circumstance asserted that the hydrophilic chitin exhibited a transition of hydrophilic character to the hydrophobic nature due to the rise of hydrogen bonds contributed by the acetyl groups or decrease of DD value. Further decrease of DD (<50%) influenced the hydrophobic nature of the polymer chains and three different size distributions were achieved including two previous size distribution. The aggregation and agglomeration of chitin chains increased the size distribution to micrometer sized (average diameter). It indicated the development of complete hydrophobic bonds influenced by higher amount of acetyl groups than the higher DD (>50%). Chitin structure at low DD (<50%) contains the largest amount of N-acetyl-D-glucosamine units and exhibits distinct domains of crystallinity. Therefore, the fraction of N-acetyl-D-glucosamine units has high influence on the solubility and solution property.

3.2. Effect of solution pH

The aggregation behavior of chitosan is strongly influenced by the pH of the solvent medium. In general, chitosan molecules are more or less ionized up to pH 6.0, and the ionization increases as the pH moves to low values. Therefore, the amino groups of chitin chains (low DA) at a particular pH (<6.0) capture H^+ solution ions and exhibited positive surface charge which can be determined by zeta potential value. The charged amino groups resist the aggregation of chitosan in the solution, but when the DA value increases from zero to higher value the aggregation starts to dominate over the coulombic repulsion forces of the charged groups. The pH at which the net charge of a chitosan solution prevents aggregation is called critical pH. Therefore, the aggregation behavior of chitosan can be divided into two distinct types—closed and open type aggregation based on pH values. The closed type aggregation is observed at very low pH when the radius of gyration becomes constant and insensitive to chitosan concentration. This indicates that chitosan is completely protonated and the solubilized chitosan molecules maintain stable aggregation. This phenomenon can be related to the classical Rayleigh theory in all charged species fragmented into smaller charged species beyond a certain critical point of net charge. On the other hand, open type aggregation takes place when aggregation and association forces overcome the repulsive effect at pH higher than the critical pH. The radius of gyration increases with the incorporation of more chitosan molecules in the solution system. In addition, the high pH value (>6.0) also raises the number of deprotonated species in the solution and aggregation moves to agglomeration due to the generation of hydrogen bonds involving the neutralized NH_2 groups of chitosan chains [8]. Moreover, the hydrolytic cleavage occurs when chitin is treated by strong acid such as highly concentrated acetic acid or HCl by involving the glycosidic bond (**Figure 2**: hydrolytic degradation). The detail mechanism has been discussed in Section 4.1.

3.3. Effect of molecular weight

Apart from the degree of deacetylation and pH, molecular weight influences the conformational changes and solubility of chitosan. Chitosan solubility increases with the decrease of molecular weight [9]. The solubilization process of chitosan, as it happens for functionalized polymers, involves different types of chemical and physical interactions such as hydrogen bonds, hydrophobic interactions, van der Waals forces, etc. The effect of DA on the

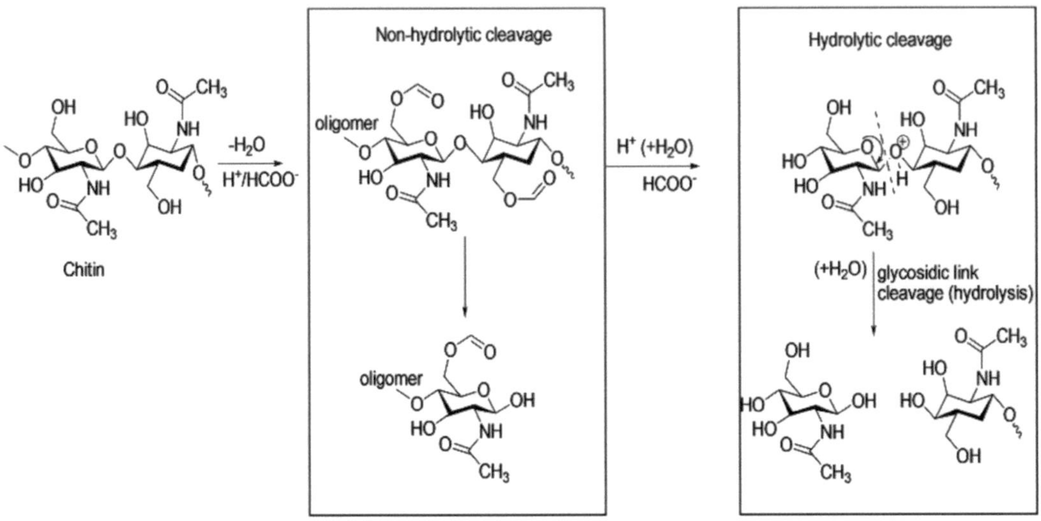

Figure 2. Mechanism of hylytic and non-hydrolytic degradation by organic acid (FA).

solubility of chitosan has already been discussed in previous section in which the hydrogen bonds involving acetyl groups played a dominant role. When DA value is lower than 50%, the protonated amino groups dominate the electrostatic repulsions between chitin chain and the hydrogen bonds collapse. As a result, chitin with low DA (<50%) become soluble at acidic pH. Chitin with low DA is fully ionized at pH 3.0 while deprotonation reaches to the highest value at pH 6.0 and precipitate occurs. Therefore, a transition between dissolved and undissolved chitin is mainly controlled by the medium pH and the degree of deacetylation or number of amino groups present in the chitin structure. The formation of hydrogen bonds and the impact of hydrophobic interaction on the chain aggregation are observed even though chitin molecules are fully deacetylated [6]. Therefore, deprotonation, $\beta \rightarrow \alpha$ phase transition and precipitation represents the scheme for the formulation of α-chitin by aggregation. Chitosan at high molecular weight (MW 300 kDa) exhibit the α-chitin crystalline structure upon aggregation. Aggregation determines a conformational entropy loss due to the arrangement of the molecular chains in a regular crystalline array [10]. The release of water compensates the loss of entropy during chain aggregation resulting in an overall decrease of Gibbs free energy, which is a thermodynamic criterion for a process to be spontaneous. However, the circumstance changes for chitosan oligomers of low MW (2.43 kDa). In this case, the aggregation is not favored due to the formation of shorter chains which reduce the hydrogen bonds formation between macromolecules and lack of amino groups for the formation of intermolecular hydrogen bonds. As a result, the pH for the transition between dissolved and undissolved chitosan in aqueous medium shifted to pH 8.0 from pH 6.0. Therefore, the soluble and insoluble transition of chitosan occurs when the MW (weight average) range exists in the range of 3.82–4.67 kDa. The transition moves to complete solubility when the MW decreases to the monomer scale at below 3.28 kDa because no intermolecular hydrogen bond leads to chitosan aggregation and solubilization is observed at neutral pH [9].

3.4. Salt effect/ionic strength and temperature

Ionic strength is a measure of the total concentration of ions present in a solution. In general, charged particles exhibit a net electrostatic effect up to a distance in the solution which is indicated as Debye screening length (k^{-1}). The electric field affects the electrokinetic phenomena and migration of the charged colloidal particles in the solution. As a result, a double layer or boundary layer is developed in chitosan solution due to the total polysaccharide charged particles and the counterions around the particles. Debye screening length (k^{-1}) can be calculated as a function of ionic strength (I) using Eq. (1),

$$k^{-1} = \sqrt{\frac{1000\varepsilon\, k_b\, T}{8\pi\, e^2 N_{av} I}} \tag{1}$$

where the parameter ε indicates water permittivity, k_b is to Boltzmann constant, T is the temperature (K), e is the electronic charge (coulombic) and N_{av} is the Avogadro number (6.022×10^{23}). It is observed that k^{-1} and I are inversely related to each other; however, the relationship is not linear but expressed by the square root. Chitosan acts as polyelectrolyte, when it is dissolved in a solvent. The polycationic solution develops electrostatic repulsive interactions while masking other available interactions. The addition of a salt or the increase of ionic strength from 9×10^{-3} to 0.46 M of the chitosan solution results in an inversion from repulsive to attractive interaction and the k^{-1} value decreases from 3.0 to 0.45 nm. The attraction inspired by the screening of amino-charged chitosan chains with anions such as CH_3COO^- and Cl^- increases the tendency of flocculation or precipitation of chitosan. Therefore, the increase of ionic strength (>0.46 M) enhances the aggregation by chitosan-chitosan attraction over the chitosan-solvent interaction and influences chitosan solubility [11]. Chitosan in acidic medium shows an expanded conformation structure since the amino groups exert repulsive force with each other, but the addition of salt or increase of ionic strength shrink the structure by increasing chain flexibility. As a result, the occupied volume of chitosan chains in solution is reduced by increasing the ionic strength and a decrease of intrinsic viscosity of chitosan solution is observed. The intrinsic viscosities of chitosan solution decreased from 2.9 to 0.71 L/g for the increase of ionic strength from 9×10^{-3} to 0.46 M. Temperature dominates the formation and dissolution of hydrogen bonds between acetyl and hydroxyl groups up to a threshold limit of ionic strength. The temperature at which the dissolution occurs is called dissolution temperature and can be defined as a function of ionic strength above the threshold value of ionic strength (i.e. critical ionic strength, I_c). Above I_c, the dissolution temperature is proportional to the ionic strength, therefore the dissolution temperature increases as the ionic strength of the chitosan solution increases [12].

4. Solvability of different solvents

4.1. Organic acids as solvent

Chitosan or modified chitin is readily soluble in dilute acidic medium below its pKa (pH = 6.5) while chitin is insoluble in organic and regular solvents. The amino groups in chitosan backbone enhance ionization at low pH by forming chit-NH_3^+ and increase the solubility of the polysaccharide while at higher pH value (>pH 6.0), it precipitates. Therefore, the solubility of chitin can be increased by converting to chitosan (by deacetylation reaction) which depends on the pKa value and also on the DD. The ability of acidic media to protonate chitosan mainly controls the ionization and solubility of the polyelectrolytes. Chitosan exhibits solubility in acid media (1%) such as acetic acid, formic acid [13], L-glutamic acid, lactic acid, succinic acid [14], etc. Tsao and coworkers proposed a mechanism for depolymerization of chitosan in acetic acid. The process involves two main pathways by chain scission, that is, (i) the depolymerization of glycosidic linkage by hydrolysis and (ii) the deacetylation of the *N*-acetyl groups. In the initial stage of dissolution, the protonation of glycosidic oxygen atom occurs at C—O bonds and form a conjugated acid. The process of protonation involves the cleavage of the exocyclic part of glucosamine structure from O-5 to C1. In this step, the cleavage of the glycosidic linkage may follow two distinct paths based on two conformations, that is, (i) chair (high energy) and (ii) half-chair (low energy) which are related to the required energy of proton transfer. Since the oxonium ion resulting from the protonation of both distorted conformation is unstable, a small energy pathway is observed to form oxocarbenium ion followed by the half-chair conformation [15]. The water molecules in the form of H_3O^+ due to a nucleophilic attack in protonation, lead to the formation of reducing sugar and these hydrolysis products are readily soluble in water. At the end of the depolymerization reaction, the resulted products are monomers, acetic acids and some other molecules (**Figure 2**: hydrolytic cleavage).

4.1.1. Formic acid (FA)

Chitin liquefaction is one of the simple processes in which chitin is transformed into small soluble molecules. Formic acid (FA) can be used as liquefaction agent for chitin. Moreover, due to high vapor pressure FA can be evaporated without leaving any residue. Three different types of products are formed after the liquefaction of chitin, that is, (i) N-acetyl glucosamine having formate functional groups (NAGF), (ii) dehydrated N-acetyl glucosamine (DH) and (iii) 5-(formyloxymethyl)furfural (FMF). The total yield in this process was achieved around 16.1% in which the highest fraction of yield was found at 10.5% for N-acetyl glucosamine having formate functional groups. In addition, the dehydrated products are achieved around 3.6% while the FMF fractions are significantly lower around 2.0%. The yield of these end products depends on the time and temperature of the liquefaction process. For example, the total yield is increased to 60% (i.e. NAGF 32.7%, DH 11.3% and FMF 16.0%) when the temperature is raised to 100°C and keeping other parameters constant. In contrast, the total yields of 28% (i.e. NAGF 12.7%, DH 11.1% and FMF 4.7%) and 57.8% (i.e. NAGF 13.2%, DH 10.0% and FMF 34.6%) are achieved after only increasing the reaction times (from 12 h) to 24 and 168 h, respectively. [16]. The breakage of glycosidic linkages does not occur in the presence of strong acidic solution, due to insufficient amount of water in the reaction. Therefore, the first step of the reaction pathway follows the generation of monomers and oligomers in the form of soluble chitin by the modification of hydroxyl groups and followed by non-hydrolytic cleavage. As the water concentration increases and reaction proceeds, the depolymerization kinetics increases with the higher supply of water leading to hydrolysis and liquefaction reaction **(Figure 2)**. Anhydrous formic acid can also be as a solvent for chitin [13].

4.2. Inorganic solvents

Many inorganic acids, bases and salts are used for the dissolution of chitin and chitosan. The extensive decomposition and deacetylation of chitin can be obtained by alkali treatment, which increases the solubility in water of the regenerated chitin. The alkali chitin solution is prepared by using 10 times more alkali than chitin. The precipitation of chitin occurs by pouring the solution into acetone followed by neutralization with HCl [17]. The obtained precipitates are insoluble in water, but after 104 h of reaction the alkali treatment allows to reach the aqueous solubility. The enhanced solubility was due to the cleavage of chitin chain and led to the destruction of the crystalline structure of chitin. A prolonged treatment with NaOH also increases the degree of deacetylation up to 90% while 50% and more deacetylated chitin is dissolved in water. Einbu and coworkers analyzed the random degradation of chitin in 2.77 M NaOH and observed random coil conformation of chitin chains regardless of molecular weight [18]. Depolymerization, deacetylation and stability of chitin solution can be enhanced when urea is added to the alkali medium. Hu et al. dissolved chitosan in a mixture continuing 8 wt% NaOH and 4 wt% urea at the temperature of −20°C and stirred for 36 h. Chitin solubilization was not possible below 4 wt% NaOH and solution instability arises upon increasing the alkali solution above 12% NaOH. The addition of 2–8% urea in the 6–10% NaOH increases the solubility and stability of depolymerized chitin fragments in the solution. The explanations behind the achieved solubility and stability is the destruction of inter- and intramolecular hydrogen bonds and the role of urea is to limit aggregation leading to the stability of the solution [19]. The entire process of chitin solubilization is also dependent on the freezing temperature of chitin in a particular mixture of solvent. For example, the chitin solution in 8 wt% NaOH and 4 wt% urea exhibits the freezing point at −19°C. The presence of more than 4 wt% NaOH enables water molecules to get access into the chitin chain matrix; water is expanded and separated from the NaOH molecules at temperature below freezing point. The volume expansion of the chitin matrix upon freezing-induced stretch and collapses of the hydrogen bonds, which brought to depolymerization and solubilization of chitin chains (**Figure 3**). In contrast, the extent of chitin deacetylation in the alkali solution was greater than the same process in presence of urea. The DA value reduced from 94 to 84% after 480 h storage in the mixture of 8 wt% NaOH and 4 wt% urea indicating that urea stabilized the chitin solution and stopped the deacetylation process. Another similar study was carried out by Fang and coworkers who described the insight mechanism of dissolution property of chitin in NaOH-urea mixture [20]. The combined system (NaOH/urea) was quite suitable to prepare a chitin solution at −30°C. The hydrated NaOH captured the chitin chains by hydrogen bonds and formed complexes while urea clusters surrounded outside the complexes as a shell-like structure. The chitin chains were separated by the hydrated NaOH and urea disrupting the inter- and intramolecular hydrogen bonding network and displayed a complete dissolution. The solution was sensitive to temperature and concentration and formed an extended worm-like structure confirmed by TEM, AFM and DLS analysis [21]. Gong and coworkers have already reported a study recently with KOH and KOH-urea as a solvent for chitin dissolution. The chitin solubility was around 80% in the aqueous KOH solution (8.4–25 wt%) and the dissolution power of bases was in the order KOH > NaOH > LiOH at −30°C. Importantly, the degree of acetylation decreased only 12.5% after the treatment with KOH and storage at 4°C

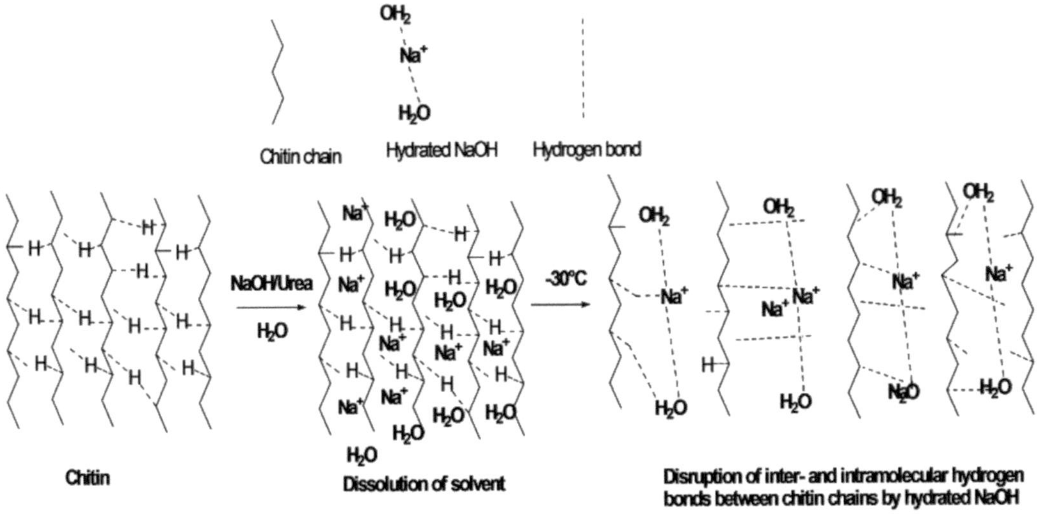

Figure 3. Hydrolysis by NaOH.

for 15 days. Urea did not exhibit any significant effect for enhancing the solvent capability of KOH [21]. Moreover, chitin solubility was also observed in a mixture of 5% LiCl and N,N-dimethylacetamide (DMA). The solution obtained from the mixture solubilized only 2 wt% chitin at 120°C and produced a gel, but 3 wt% chitin was dissolved when lithium thiocyanate was used as a solvent at 100°C [22]. Also, the recovery of the product in a strong acidic environment is quite difficult and expensive. For example, the hydrolysis of chitin/chitosan or depolymerization by 12.07 M inorganic acid (HCl) at 40°C for 28 h does not produce any N-deacetylated moiety [23]. Therefore, use of nitrous acid is a cheaper alternative for chitosan depolymerization by the cleavage of glycosidic bonds but the stoichiometry of the reaction depends on the amount of acidic solution. Water soluble chitosan oligosaccharides and highly deacetylated chitosan oligosaccharides were synthesized using nitrous acid [24]. The depolymerization process was carried out by adding sodium nitrite ($NaNO_2$) to the chitosan solution (2% acetic acid) and keeping the solution for 3 h at pH 7.0, then the excess water in the reaction was evaporated at 50°C. The fractionation and extraction were carried out by methanol and filtrated for separation. The depolymerization started with the deamination of the 2-amino-2-deoxy-β-D-glucopyranose ($GlcNH_2$) units and produced a new reducing sugar of 2,5-anhydro-D-mannose (M-units). The M-units did not exhibit any interconversion between anomers of the reducing sugar. The aldehyde groups of the M-unit were quite reductive to amino groups by avoiding any intramolecular hemiacetal formation **(Figure 4)**. Therefore, two M-units form imino linkage from the reaction between NH_2 and –CHO by Schiff base reaction at pH > 5.0 and undergoes several steps of water elimination in acidic medium. As a result, 5-hydroxymethyl-2-furfural (HMF), depolymerized and fully deacetylated chitosan were achieved. The depolymerization process provided water-soluble chitosan of different molecular weights, which can be collected from various methanol fractions. Some solvent systems for chitin have been reviewed by Pillai and coworkers [25]. They reviewed some solvent systems from previous works, highlighting that LiCl-tertiary amide solvent system can dissolve 5% (w/v) chitin. Moreover, LiCl also forms a coordination complex with chitin which is dissolved in dimethylacetamide (DMA) and N-methyl-2-pyrrolidone (NMP). Calcium chloride-methanol system as a solvent for chitin was also reported in the review. Chitin can form a complex with calcium and dissolve in methanol but the limitation arises due to very high viscosity, which makes the scale up difficult.

4.3. Ionic liquids (ILs) as solvent

Strong organic and inorganic acids, strong alkali solution or other inorganic solvents such as LiCl-tertiary solvent, $CaCl_2$-MetOH system possess some disadvantages like corrosiveness, volatility, toxicity and so on. Moreover, inappropriate segmentation of chains, unstable yields occur during the hydrolysis or depolymerization in those solvents. As a suitable alternative, ionic liquids (ILs) are considered as green solvents due to their non-volatility, excellent solvation power, wide temperature ranges in the liquid phase, strong polarity and stability of end products. 1-butyl-3-methylimidazolium chloride (BminCl) is an ionic liquid (IL) which gives a swelled state of 5 wt% chitin after treatment at 130°C for 5 h [3]. The swelling of chitin in the IL, [BminCl] occurs due to the strong coordination of the Cl^- ions and partially break the hydrogen bonds of chitin chains. The complete solubility of chitin is only possible when a stronger coordinating anion than Cl^- ion will destroy the entire hydrogen bonds network (—NH···O═C and —OH···O═C) produced by *N*-acetyl groups. Therefore, 1-butyl-3-methylimidazolium acetate (BminAc) was used as solvent and found better solubility than the (BminCl). The acetate ions in (BminAc) exhibit itself as a strong

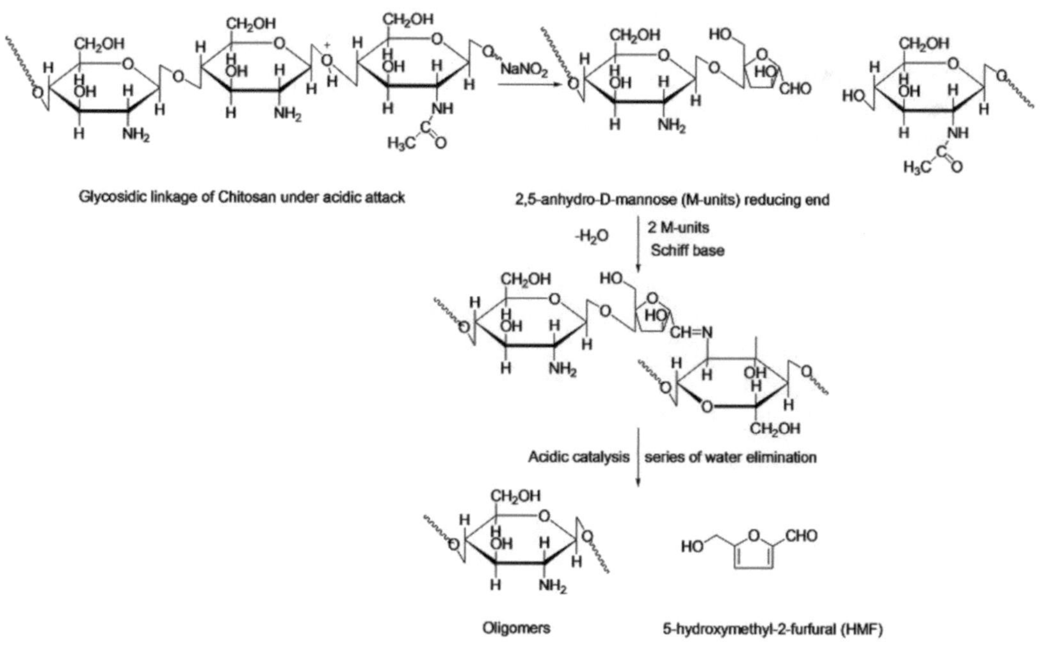

Figure 4. Hydrolysis by $NaNO_2$.

conjugate base of a weak acid, which can interact with the H-bonds of chitin. It destroys the hydrogen bonds and solubilizes the crystal chitin. Xie and coworkers investigated the IL, 1-butyl-3-methyl-imidazolium chloride ([Bmim]Cl) as a solvent but they achieved partial dissolution of chitin and chitosan at 110°C probably due to the moderated polarity of the IL [26]. Prasad and coworkers found that 1-allyl-3-methylimidazolium bromide exhibited good solvent property for the dissolution of 5 wt% chitin when it was heated at 100°C for 48 h [27]. The obtained chitin solution was quite clear and homogeneous as confirmed by scanning electron microscopy analysis. The chitin powder recovered and regenerated by methanol treatment possessed the same crystalline structure as that of the crab shell. No degradation of chain or decrease of molecular weight had been occurred. Upon increasing the chitin amount to 7 wt%, a gel was obtained. Wang and coworkers studied the effect of three ILs: alkyl imidazolium chloride ([AMIM]Cl), alkyl imidazolium dimethyl phosphate ([MMIM][Me_2PO_4]) and 1-allyl-3-methyl-imidazolium acetate ([AMIM]Ac) on the solubility of a series of different DA (degree of acetylation) of chitin [28]. The dissolution of 5 wt% chitin occurs at 110°C in the [AMIM]Ac while [AMIM]Cl and [MMIM][Me_2PO_4] can dissolve 0.5 wt% and 1.5 wt% chitin at 45 and 60°C, respectively. The observation showed that acetate anions are more efficient to break down the network of hydrogen bonds than the chloride and dimethyl phosphate anion. Similar findings were observed when 1-butyl-3-methylimidazolium acetate ([BMIN]Ac) and 1-butyl-3-methylimidazolium chloride ([BMIN]Cl) were used as solvent at room temperature [3]. The study showed that limited chitin solubility (1 wt%) was achieved in [BMIN]Cl while the solubility increased to 5 wt% in [BMIN]Ac at room temperature. Qin and coworkers reported on the dissolution of chitin using IL, 1-ethyl-3-methylimidazolium acetate for which the required temperature was 100°C while the dissolution of cellulose with the same IL was obtained at 40°C [26]. The difference was due to the structural arrangement of chitin with acetamide group on the C2 position while cellulose shows a hydroxyl group in the same position. The major obstacle to dissolution was the strong hydrogen bonds between C = O and NH groups of the adjacent chitin chains distributed in the chitin cluster. Thanks to their strong polarity, ILs overcome the energy barrier and makes it possible to dissolve in the molten state (40°C). Shimo and coworkers used tris(2-hydroxyethyl)methylammonium (THEMA) type ILs in the absence and presence of ethylenediamine (EDA) to dissolve chitin at mild non-aqueous conditions [29]. Four different THEMA-type ILs were used to dissolve chitin: Tris(2-hydroxyethyl)methylammonium acetate ([THEMA][OAc]), Tris(2-hydroxyethyl) methylammonium methyl sulfate ([THEMA][MeOSO3]), Tris(2-hydroxyethyl) methylammoniumtrifluoromethansulfonate ([THEMA][CF3SO3]). Partial and total solubility was observed when THEMA-type ILs used as solvents in the absence and presence of EDA, respectively. For example, ([THEMA][OAc]) exhibited excellent solubility at room temperature in the presence of EDA. The reason behind the complete dissolution was revealed with the help of X-ray diffraction analysis. The analysis exhibited that the EDA penetrated into the crystalline α-chitin and formed a complex. Therefore, when EDA was added to the system of IL and chitin, the EDA easily broke the hydrogen bonds present in the α-chitin and created strong hydrogen bonds of with the IL. This mechanism leads to dissolving the dissolution of chitin at room temperature by loosen their inter-chain hydrogen bonds between chitin chains. Some references about the use of ILs in the modification of chitin and chitosan to enhance the solubility are given in **Table 1**.

Chitin/chitosan	Solvent	References
Acetylated chitin	Dimethyl sulfoxide (DMSO)	[30]
Chitin-graft-polystyrene	Dimethyl sulfoxide (DMSO)	[31]
Monomethyl-modified chitosan	Water	[32]
O-alkylated chitosan	Chloroform, ethanol, water and acetic acid	[33]
Chitosan-graft-polycaprolactone	Dimethylformamide (DMF), DMSO, ethanol and toluene	[34]

Table 1. Modified chitin and chitosan with different ionic liquids and solubility in solvents.

4.4. Enzymatic hydrolysis

Enzymatic hydrolysis is a green process to achieve chitosan excellent solubility in water by producing chitooligosaccharides (COS). The process does not require extreme conditions (very low pH or high concentration of acids) and the tuning of molecular weight, and DD of final product can be achieved by avoiding any unwanted yield. Chitinase, chitosanase are specific enzymes and many other nonspecific enzymes such as glycanases, proteases, lipase are isolated from many biological sources. Unlike acid hydrolysis, enzymatic hydrolysis affects both the depolymerization and deacetylation of chitin or chitosan through the catalytic activities, which mainly depends on the molecular structure of chitinase (enzyme). Chitinase contains four catalytic domains in its structure, that is, (i) signal sequence, (ii) catalytic domain, (iii) serine/threonine region which can accept O-glucosylation and (iv) C-terminal chitin-binding domain [41]. The depolymerization occurs in the similar style of classical acid-base catalytic reaction (**Figure 3**: hydrolytic cleavage) followed by retention (two steps) or inversion (one step) reaction. In the retention mechanism, firstly, the acidic residues release protons, cleavage of glycosidic bonds and subsequently positively charged oxocarbonium ions intermediates are produced. Then, secondly, the intermediates are stabilized with the help of intermediate covalent bonds (glycosal-enzyme) but the subsequent reaction with the water molecules leads to the retention of anomeric configuration again. However, the inversion mechanism involves negative-charged residue, carbonium intermediate and water molecules at a time for the degradation of chain and inversion of the anomeric configuration. Moreover, the deacetylation of chitin molecules takes place by deacetylase treatment. The chitin deacetylase enzyme isolates from *Mucor rouxii* and forms enzyme-polymer complexes with chitin. The hydrolysis of acetyl groups occurs by the enzymatic attack to three acetyl groups (maximum) before the reaction proceeds to next complex formation. The enzymatic deacetylation yields randomly distributed GlcNAc and GlcN residues in the form of block copolymers. Pronase was reported to increase the solubility and decrease the degree of acetylation (DA) [42]. The chitosan solution in 1% acetic acid was treated with pronase at 100:1 ratio at 37°C for 1–5 h with subsequent and re-precipitation by using NaOH (2 M). The treated chitosan exhibited the molecular weight of 8.5 kDa while the native chitosan was 71 kDa. Moreover, the DA value decreased from 25 to 14% after the isolation of new chitosan. Consequently, the chitosan after pronase digestion can be solubilized around 66–74 wt% in 0.01% acetic acid which was 13 wt% for the raw chitosan. The other enzymes used for chitosan enzymatic treatments and the operation conditions have been summarized in **Table 2**.

Enzyme	Water soluble modified chitosan	References
Lysozyme	Chitooligosaccharides (COS)	[35, 36]
Papain		
Cellulase		
O-glycoside hydrolase (EC 3.2.1)	Low MW chitosan (3–6 kDa)	[37]
Chitosanase and β-D-glucosaminidase	D-glucosamine	[38]
Chitin deacetylases	Chitin and chitosan oligomers	[39]
Carbohydrases from Myceliophthora sp	Low MW chitosan (4–28 kDa with 85% DD)	[40]

Table 2. List of enzymes for the hydrolysis of polysaccharide (chitin, chitosan).

5. Modified chitosan

Modification of molecular structure can enhance the solubility of chitosan in water. Phosphorylated chitosan, quaternized chitosan derivatives and carboxymethyl chitosan can be solubilized

Modified chitosan	Solubility in	References
Phosphorylated chitosan		
N-methylene phosphonic chitosan	Water, HCl, acetic acid	[43]
Chitosan diethyl phosphate	Diluted organic or mineral acid	[44]
α-Galactosyl-chitosan conjugates	Water	[45]
Chitosan-dendrimer hybrid	Water	[46]
Quaternized chitosan derivatives		
N-phenmethyl-N,N-dimethyl chitosan (PDCS) N-(1-pyridylmethyl-2-ylmethyl)-N,N-dimethyl chitosan N-(1-pyridylmethyl-3-ylmethyl)-N,N-dimethyl chitosan N-(1-pyridylmethyl-4-ylmethyl)-N,N-dimethyl chitosan	0.1–1.6 mg/ml in water	[47]
N-[(2-hydroxy-3trimethylammonium)propyl]chitosan chloride (HTACC)	Water	[48]
Carboxymethyl chitosan		
O-carboxymethyl chitosan sodium	N-methylmorpholine-N-oxide (NMMO)	[49]
N,N-dicarboxymethyl chitosan	Water	[50]
N,O-carboxymethyl chitosan	Water	
Photosensitive chitosan with benzene group	Benzene and toluene	[5, 51]
Dibutyryl chitin	DMF, DMSO, dimethylacetamide (DMAc) and ethanol	[52]

Table 3. Modified chitosan and solubility in different solvents.

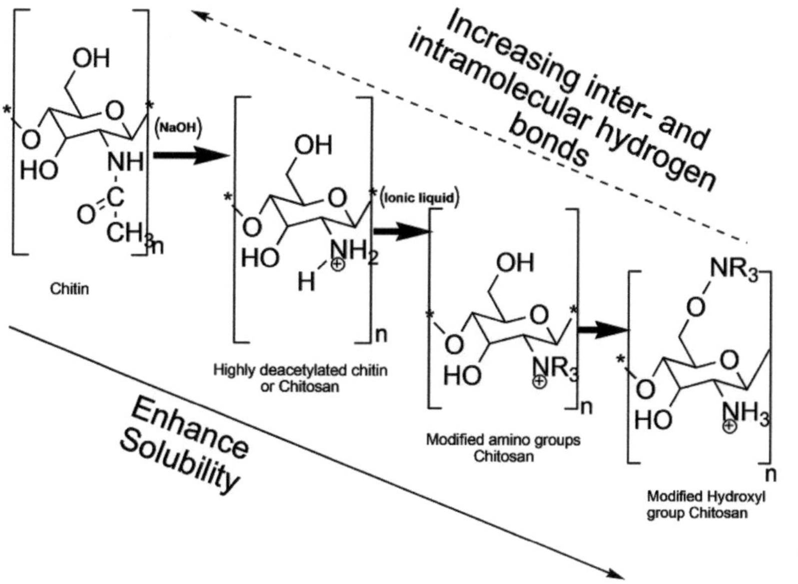

Figure 5. Trend of solubility and intermolecular hydrogen bond between acetyl groups.

in different solvents at ambient conditions (**Table 3**). The solubility trend of chitin based on the modification of the molecular structure has been clearly displayed in **Figure 5**.

6. Conclusion and future perspectives

Chitin and chitosan have shown a big potential in pharmaceuticals, biomedical, agricultural sectors as well as food and textiles industry. Despite the myriad of opportunities, chitin and chitosan poor solubility in the most common solvents is a greatly limitation for scaling up the process from lab to industrial level. High viscosity of chitin and chitosan solution is another drawback with great impact of processing operations and equipment requirements. Even though chitin is sparingly soluble in strong acidic solution, corrosive and hazardous solvents should not be practiced to meet up regulatory compliances concerning chemical safety management. In conclusion, much works is still required to exploit the opportunities of this futuristic material which can contribute to economically feasible industrial growth.

Acknowledgements

The project work has been funded by EU under the framework of Erasmus Mundus Program in Sustainable Management and Design for Textile (SMDTex).

Author details

Jagadish C. Roy[1,2,3,4*], Fabien Salaün[1,2], Stéphane Giraud[1,2], Ada Ferri[3], Guoqiang Chen[4] and Jinping Guan[4]

*Address all correspondence to: jagadish-chandra.roy@ensait.fr

1 University Lille Nord de France, Lille, France

2 ENSAIT, GMTEX Lab, Roubaix, France

3 Departments of Applied Science and Technology (DISAT), Polytecnico di Torino, Italy

4 College of Textile and Clothing Engineering, Soochow University, Suzhou, Jiangsu, China

References

[1] Austin PR. Solvents for and Purification of Chitin. United States Patent1973. p. 2-4

[2] Agboh OC, Qin Y. Chitin and chitosan fibers. Polymers for Advanced Technologies. 1997;**8**:355-365

[3] Wu Y, Sasaki T, Irie S, Sakurai K. A novel biomass-ionic liquid platform for the utilization of native chitin. Polymer. 2008;**49**(9):2321-2327

[4] Rinaudo M. Chitin and chitosan: Properties and applications. Progress in Polymer Science (Oxford). 2006;**31**(7):603-632

[5] Guiping M, Yang L, Kennedy JF, Nie J. Synthesize and properties of photosensitive organic solvent soluble acylated chitosan derivatives (2). Carbohydrate Polymers. 2011;**84**:681-685

[6] Philippova OE, Korchagina EV, Volkov EV, Smirnov VA, Khokhlov AR, Rinaudo M. Aggregation of some water-soluble derivatives of chitin in aqueous solutions: Role of the degree of acetylation and effect of hydrogen bond breaker. Carbohydrate Polymers. 2012;**87**(1):687-694

[7] Popa-Nita S, Alcouffe P, Rochas C, David L, Domard A. Continuum of structural organization from chitosan solutions to derived physical forms. Biomacromolecules. 2010;**11**(1):6-12

[8] Philippova OE, Korchagina EV. Chitosan and its hydrophobic derivatives: Preparation and aggregation in dilute aqueous solutions 1. Polymer Science Series A. 2012;**54**(7):552-572

[9] Tian M, Tan H, Li H, You C. Molecular weight dependence of structure and properties of chitosan oligomers. RSC Advances. 2015;**5**(85):69445-69452

[10] Recillas M, Silva LL, Peniche C, Goycoolea FM, Rinaudo M, Argüelles-Monal WM. Thermoresponsive behavior of chitosan-g-N-isopropylacrylamide copolymer solutions. Biomacromolecules. 2009;**10**(6):1633-1641

[11] Cho J, Heuzey M-C, Bégin A, Carreau PJ. Viscoelastic properties of chitosan solutions: Effect of concentration and ionic strength. Journal of Food Engineering. 2006;**74**(4):500-515

[12] Tanaka K, Nishida K, Gabrys BJ, Lawrence MJ, Kanaya T. Critical dissolution ionic strength of aqueous solution of chitosan hydrochloride salt. Journal of the Society of Fiber Science and Technology. 2014;**70**(9):225-231

[13] Hunt S, Huckerby TN. A comparative study of molluscan and crustacean chitin proteoglycans by carbon-13 NMR spectroscopy. Identification of carbohydrate and amino acid contributions and the determination of amino acid chemical shifts in anhydrous formic acid. Comparative Biochemistry and Physiology Part B: Comparative Biochemistry. 1987;**88**(4):1107-1116

[14] Romanazzi G, Gabler FM, Margosan D, Mackey BE, Smilanick JL. Effect of chitosan dissolved in different acids on its ability to control postharvest gray mold of table grape. Phytopathology. 2009;**99**(9):1028-1036

[15] Smith BJ. A conformational study of 2-oxanol: Insight into the role of ring distortion on enzyme-catalyzed glycosidic bond cleavage. Journal of the American Chemical Society. 1997;**119**(11):2699-2706

[16] Zhang J, Yan N. Formic acid-mediated liquefaction of chitin. Green Chemistry. 2016;**18**(18):5050-5058

[17] Sannan T, Kurita K, Iwakura Y. Studies on chitin, 1-effect of deacetylation on solubility. Die Makromolekulare Chemie. 1976;**177**(12):3589-3600

[18] Einbu A, Naess SN, Elgsaeter A, Varum KM. Solution properties of chitin in alkali. Biomacromolecules. 2004;**5**(5):2048-2054

[19] Hua X, Dua Y, Tanga Y, Wanga Q, Fenga T, Jianhong Yanga JFK. Solubility and property of chitin in NaOH/urea aqueous solution. Carbohydrate Polymers. 2007;**70**:451-458

[20] Fang Y, Duan B, Lu A, Liu M, Liu H, Xu X, Zhang L. Intermolecular interaction and the extended wormlike chain conformation of chitin in NaOH/urea aqueous solution. Biomacromolecules. 2015;**16**(4):1410-1417

[21] Gong P, Wang J, Liu B, Ru G, Feng J. Dissolution of chitin in aqueous KOH. Cellulose. 2016;**23**(3):1705-1711

[22] Gagnaire D, Saint-Germain J, Vincendon MNMR. Studies of chitin and chitin derivatives. Die Makromolekulare Chemie. 1982;**183**(3):593-601

[23] Einbu A, Grasdalen H, Varum KM. Kinetics of hydrolysis of chitin/chitosan oligomers in concentrated hydrochloric acid. Carbohydrate Research. 2007;**342**(8):1055-1062

[24] Lin CW, Lin JC. Characterization and blood coagulation evaluation of the water-soluble chitooligosaccharides prepared by a facile fractionation method. Biomacromolecules. 2003;**4**(6):1691-1697

[25] Pillai CKS, Paul W, Sharma CP. Chitin and chitosan polymers: Chemistry, solubility and fiber formation. Progress in Polymer Science (Oxford). 2009;**34**(7):641-678

[26] Qin Y, Lu X, Sun N, Rogers RD. Dissolution or extraction of crustacean shells using ionic liquids to obtain high molecular weight purified chitin and direct production of chitin films and fibers. Green Chemistry. 2010;**12**(6):968

[27] Kamalesh P, Masa-aki M, Yoshiro K, Akihiko T, Yoshifumi N, Jun-ichi K. Weak gel of chitin with ionic liquid, 1-allyl-3-methylimidazolium bromide. International Journal of Biological Macromolecules. 2009;**45**:221-225

[28] Wang WT, Zhu J, Wang XL, Huang Y, Wang YZ. Dissolution behavior of chitin in ionic liquids. Journal of Macromolecular Science, Part B. 2010;**49**(3):528-541

[29] Shimo M, Abe M, Ohno H. Functional comparison of polar ionic liquids and onium hydroxides for chitin dissolution and deacetylation to chitosan. ACS Sustainable Chemistry and Engineering. 2016;**4**(7):3722-3727

[30] Mine S, Izawa H, Kaneko Y, Kadokawa J. Acetylation of α-chitin in ionic liquids. Carbohydrate Research. 2009;**344**(16):2263-2265

[31] Yamamoto K, Yoshida S, Mine S, Kadokawa J. Synthesis of chitin-graft-polystyrene via atom transfer radical polymerization initiated from a chitin macroinitiator. Polymer Chemistry. 2013;**4**(11):3384

[32] Zhaodong W, Liuchun Z, Chuncheng L, Dong Z, Yaonan X, Guohu G, Wenxiang Z. Modification of chitosan with monomethyl fumaric acid in an ionic liquid solution. Carbohydrate Polymers. 2015;**117**:973-979

[33] Chen H, Cui S, Zhao Y, Wang B, Zhang S, Chen H, Peng X. O-alkylation of chitosan for gene delivery by using ionic liquid in an in-situ reactor. Engineering. 2012;**4**(10):114-117

[34] Yang L, Zhang J, He J, Zhang J, Gan Z. Homogeneous synthesis of amino-reserved chitosan-graft-polycaprolactone in an ionic liquid and the application in cell cultivation. Polymer International. 2015;**64**(8):1045-1052

[35] Laokuldilok T, Potivas T, Kanha N, Surawang S, Seesuriyachan P, Wangtueai S, Phimolsiripol Y, Regenstein JM. Physicochemical, antioxidant, and antimicrobial properties of chitooligosaccharides produced using three different enzyme treatments. Food Bioscience. 2017;**18**(April):28-33

[36] Zimoch-Korzycka A, Gardrat C, Castellan A, Coma V, Jarmoluk A. The use of lysozyme to prepare biologically active chitooligomers. Polímeros. 2015;**25**(1):35-41

[37] Il AV, Varlamov VP, Melent AI, Aktuganov GE. Depolymerization of chitosan with a chitinolytic complex from bacteria of the genus Bacillus sp. 739. Applied Biochemistry and Microbiology. 2001;**37**(2):142-144

[38] Sun Y, Zhang J, Wu S, Wang S. Preparation of d-glucosamine by hydrolysis of chitosan with chitosanase and β-d-glucosaminidase. International Journal of Biological Macromolecules. 2013;**61**:160-163

[39] Hamer SN, Cord-Landwehr S, Biarnés X, Planas A, Waegeman H, Moerschbacher BM, Kolkenbrock S. Enzymatic production of defined chitosan oligomers with a specific

pattern of acetylation using a combination of chitin oligosaccharide deacetylases. Scientific Reports. 2015;**5**(1):8716

[40] Khasanova LM, Il'ina AV, Varlamov VP, Sinitsyna OA, Sinitsyn AP. Hydrolysis of Chitozan with an enzyme complex from Myceliophthora sp. Applied Biochemistry and Microbiology. 2014;**50**(4):422-428

[41] Struszczyk MH. Chitin and chitosan: Part III. Some aspects of biodegradation and bioactivity. Polimery/Polymers. 2002;**47**(9):619-629

[42] Vishu Kumar AB, Varadaraj MC, Gowda LR, Tharanathan RN. Low molecular weight chitosans-preparation with the aid of pronase, characterization and their bactericidal activity towards *Bacillus cereus* and *Escherichia coli*. Biochimica et Biophysica Acta—General Subjects. 2007;**1770**(4):495-505

[43] Heras A, Rodri'guez NM, Ramos VM, Agullo E. N-methylene phosphonic chitosan: A novel soluble derivative. Carbohydrate Polymers. 2001;**44**:1-8

[44] Cardenas G, Cabrera G, Taboada E, Rinaudo M. Synthesis and characterization of critosan alkyl phosphate. Journal of the Chilean Chemical Society. 2006;**51**(1):815-820

[45] Sashiwa H, Thompson JM, Das SK, Shigemasa Y, Tripathy S, Roy R. Chemical modification of chitosan: Preparation and lectin binding properties of alpha-galactosyl-chitosan conjugates. Potential inhibitors in acute rejection following xenotransplantation. Biomacromolecules. 2000;**1**(3):303-305

[46] Sashiwa H, Yajima H, Aiba S. Synthesis of a chitosan – Dendrimer hybrid and its. Biomacromolecules. 2003;**4**:1244-1249

[47] Wei L, Li Q, Tan W, Dong F, Luan F, Guo Z. Synthesis, characterization, and the antioxidant activity of double quaternized chitosan derivatives. Molecules. 2017;**22**(3):501

[48] Noppakundilograt S, Buranagul P, Graisuwan W, Koopipat C, Kiatkamjornwong S. Modified chitosan pretreatment of polyester fabric for printing by ink jet ink. Carbohydrate Polymers. 2010;**82**:1124-1135

[49] Liu X, Wang Q, Zhuang X, Wu B, Yang F, Zeng A. Study on antibacterial activity of O-carboxymethyl chitosan sodium salt and spinnability of O-carboxymethyl chitosan sodium salt/cellulose polyblends in N-methylmorpholine-N-oxide system. Carbohydrate Polymers. 2012;**89**(1):104-110

[50] An NT, Thien DT, Dong NT, Le DP. Water-soluble N-carboxymethylchitosan derivatives: Preparation, characteristics and its application. Carbohydrate Polymers. 2009;**75**(3): 489-497

[51] Ma G, Qian B, Yang J, Hu C, Nie J. Synthesis and properties of photosensitive chitosan derivatives(1). International Journal of Biological Macromolecules. 2010;**46**(5):558-561

[52] Guo C, Zhou L, Lv J. Effects of expandable graphite and modified ammonium polyphosphate on the flame-retardant and mechanical properties of wood flour-polypropylene composites. Polymers and Polymer Composites. 2013;**21**(7):449-456